BASIC HYDROGEOLOGIC METHODS

BASIC HYDROGEOLOGIC METHODS

A FIELD AND LABORATORY MANUAL WITH MICROCOMPUTER APPLICATIONS

Frank W. Fletcher

Department of Geological and Environmental Science
Susquehanna University

TECHNOMIC
PUBLISHING CO., INC.
LANCASTER · BASEL

Basic Hydrogeologic Methods

a **TECHNOMIC** publication

Published in the Western Hemisphere by
Technomic Publishing Company, Inc.
851 New Holland Avenue, Box 3535
Lancaster, Pennsylvania 17604 U.S.A.

Distributed in the Rest of the World by
Technomic Publishing AG
Missionsstrasse 44
CH-4055 Basel, Switzerland

Printed in the United States of America
10 9 8 7 6 5 4 3 2 1

Main entry under title:
  Basic Hydrogeologic Methods: A Field and Laboratory Manual with
  Microcomputer Applications

A Technomic Publishing Company book
Bibliography: p.
Includes index p. 309

Library of Congress Catalog Card No. 96-61337
ISBN No. 1-56676-400-9

Table of Contents

Preface

THIS book was prepared in response to a need for a practical manual of common hydrogeologic methods, particularly one that provides convenient microcomputer applications. It is designed chiefly to be used as a laboratory manual in an introductory groundwater hydrology course, to complement textbook and lecture topics, but it may also serve as a handbook of investigative analytical techniques for the professional environmental scientist whose background may not include formal training in hydrogeology.

Hydrogeology is taught in colleges and universities in many forms, ranging from the strictly practical to the highly theoretical. While a sound theoretical foundation is essential for anyone who must contend with the complex problems that are the warp-and-woof of modern day hydrogeology, numerous popular texts provide this background. In contrast, the approach of this book is "how-to-do and hands-on." Its purpose is to provide clear, step-by-step instructions in many of the fundamental methods of hydrogeologic investigation. These methods include both 1) the traditional techniques of data analysis, such as mathematical computation by electronic calculator and construction of graphs by hand plotting and 2) microcomputer techniques employing electronic spreadsheets, graphing, and gridding-and-contouring software. The microcomputer methods employ commercial software such as Lotus 1-2-3, Microsoft Excel, Quattro Pro, Golden Software's GRAPHER and SURFER, and Geraghty & Miller's AQTESOLV. Although familiarity with any of the applications is helpful, the instructions in this manual assume no prior experience with them. The instructions in the Appendix explain how to create the Hydrodata Diskette which is employed in many of the microcomputer applications.

Basic Hydrogeologic Methods is divided into three sections: Groundwater

Occurrence and Movement, Groundwater Investigations, and Aquifer and Well Hydraulics. Each section begins with a brief summary of relevant terminology and principles. This introductory chapter is followed by a case study, which may be employed to provide a practical context for the hydrogeologic methods that are described in subsequent chapters. Most of the methodological exercises culuminate in an analytical product, such as data table, graph, contour map, etc., which readily serve as a focus for problem-solving activities, classroom discussions, and investigative reports. Many of the exercises present at least two investigative methods for accomplishing a particular hydrogeologic task. For example, time–drawdown graphs may be produced by a hand-plotting method or by a microcomputer method. For the professional scientist, the choice of a particular method might depend on such factors as the time available to carry out the task, the degree of accuracy required, or the availability of accessory equipment and materials. For the introductory student, it is pedagogically sound to work first through a more fundamental method (e.g., hand plotting) before advancing to a microcomputer method (e.g., spreadsheet and graphing).

I thank the many patient and dedicated students in my course Groundwater Hydrology who helped to test and refine these exercises over the past five years. It is to these young professionals that this book is dedicated.

PART 1

GROUNDWATER OCCURRENCE AND MOVEMENT

CHAPTER 1

Principles of Groundwater Occurrence and Movement

1.1 DEFINITION AND CLASSIFICATION OF GROUNDWATER

GROUNDWATER is one element of a great global system of water movement known as the hydrologic cycle (see Figure 1.1). Other major elements of this cycle include precipitation, evaporation, transpiration, surface runoff, and infiltration. As Figure 1.1 illustrates, a portion of the water that falls to the earth in the form of precipitation infiltrates downward into the subsurface. This water, which lies beneath the land surface, is known collectively as *underground* or *subsurface water.*

Underground water occurs in two relatively distinct zones (see Figure 1.2). The soil or rock directly beneath the land surface consists of openings that are filled with both air and water; this is the *unsaturated zone.* The underlying region, in which the openings are filled entirely with water, is referred to as the *saturated zone.* The name "groundwater" is appropriately given only to the water contained in the saturated zone.

The surface that marks the top of the saturated zone is called the *water table.* The water table is defined as "the surface on which the fluid pressure p in the pores of a porous medium is exactly atmospheric" (Freeze and Cherry, p. 39). It is marked by the water level of inactive (i.e., non-pumping) wells. [For additional information, see Driscoll (1986), pp. 46–61 and Heath (1987), pp. 1 and 5.]

1.2 AQUIFERS AND CONFINING UNITS

Although groundwater fills the openings of the saturated zone, not all rocks will yield enough water to wells to be practically or economically

3

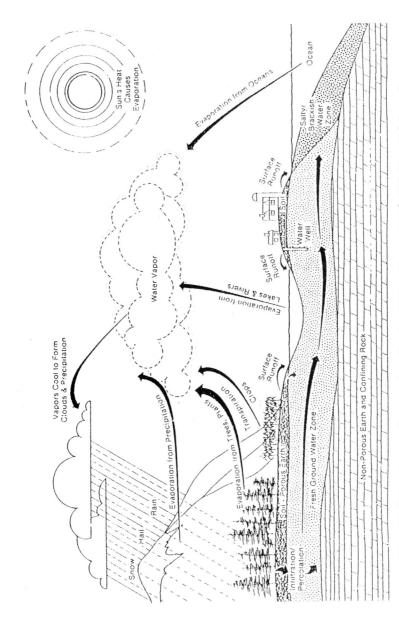

Figure 1.1 Hydrologic cycle. (Reprinted from Chandler, 1990, p. 45; copyright © 1990, *Water Well Journal*.)

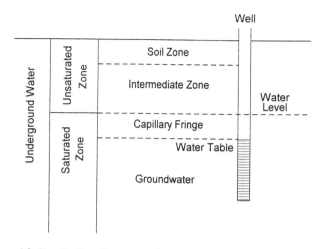

Figure 1.2 Classification of underground water (modified from Heath, 1987, p. 4).

usable. A rock unit that "contains sufficient saturated permeable material to yield significant quantities of water to wells and springs" is called an *aquifer* (Lohman et al., 1972, p. 2). Examples of aquifers include gravel, sand, and fractured or cavernous limestone. A rock that does not permit groundwater to flow through it readily (although it may store a large volume of groundwater) and restricts the movement of water into or out of aquifers is referred to as a *confining unit*. Examples of confining units include clay, shale, and unfractured basalt.

Aquifers are classified as unconfined or confined (see Figure 1.3). An unconfined or water table aquifer is one in which the groundwater is "in direct contact with the atmosphere through open spaces in permeable material" (Davis and DeWiest, 1966, p. 43). In an unconfined aquifer the water table is free to rise and fall in response to changes in the volume of groundwater in storage. A confined or artesian aquifer is one that is bounded from above by confining units. In the confined aquifer, the groundwater is under a significantly greater pressure than atmospheric pressure. In tightly cased wells open to a confined aquifer, the water level stands at some height above the top of the aquifer and marks the position of the potentiometric surface. [For additional information, see Driscoll (1986), pp. 61–66 and Heath (1987), p. 6.]

1.3 DEPTH TO GROUNDWATER

Determining "depth to groundwater" is one of the first steps of many hydrogeologic investigations. For example, "depth to water table" is an

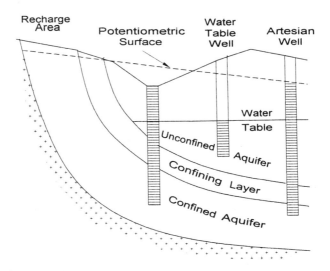

Figure 1.3 Types of aquifers (modified from Todd, 1980, p. 42).

important consideration in investigations of groundwater contamination, in the design and construction of waste disposal sites, and in the development of water supplies. A shallow water table can cause the saturation of rock and soil close to the land surface, thereby increasing the potential for groundwater contamination from waste disposal sites or leaking underground storage tanks. In contrast, a deep water table increases the depth to which monitoring or water supply wells must be drilled, and thus the cost of drilling, construction, and pumping of such wells.

Depth to water table (or depth to potentiometric surface in confined aquifers) is determined by measuring the depth to water level in wells (see Figure 1.4). If the position of the water level in the well is measured from the top of the well casing, then the depth to water table is equal to the depth to water level in the well less the height of the casing above the land surface.

Depth to water table (or potentiometric surface) may be illustrated on a *depth to water map* (see Figure 1.5). The contour lines on a depth to water map mark the vertical distance between the land surface and the water table or potentiometric surface. For unconfined aquifers a depth to water table map is similar to an isopach map, in which the difference between the land surface and the water table represents the thickness of the unsaturated zone. [For additional information, see Driscoll (1986), pp. 547–552 and Heath (1987), pp. 72–73.]

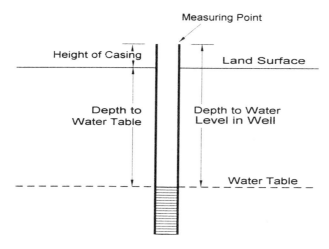

Figure 1.4 Illustration of depth to water.

1.4 WATER LEVEL FLUCTUATIONS

Groundwater levels fluctuate over time in response to various natural and human factors, including changes in the rates of precipitation, evapotranspiration, and pumping. An understanding of these fluctuations is important in the design of water supply and groundwater monitoring

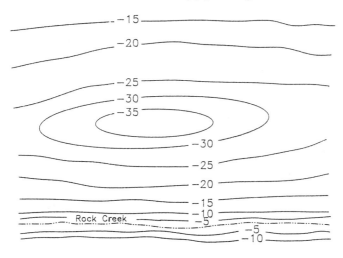

Figure 1.5 Example of a depth to water map. Contour lines represent the depth of the water table below the land surface.

programs, and particularly in the planning of the depth of the wells and the length of well screens.

Water levels in wells fluctuate continuously in response to changes in groundwater storage and may rise or fall from less than an inch to many feet in a relatively short period of time. Under natural conditions, water levels of unconfined aquifers typically exhibit a wider range of fluctuation than do those of confined aquifers; they are influenced markedly by direct recharge from precipitation, by the loss to evaporation, and the discharge to springs and streams. Even changes in atmospheric pressure will cause small fluctuations of groundwater levels in wells.

Water level fluctuations of a single well may be illustrated by means of a hydrograph. A hydrograph is a time-series graph on which time is scaled on the horizontal axis and elevation (or depth) is scaled on the vertical axis. Depending on the particular time scale selected, hydrographs can be used to display the fluctuation of the groundwater level in response to various kinds of changes in the hydrologic regime. Figure 1.6 illustrates a short-term response of the water table to the effects of a single storm. Figure 1.7 depicts a hydrograph that reflects the seasonal distribution of precipitation and evapotranspiration. Long-term hydrographs, such as those shown in Figure 1.8, reveal changes in a region's water budget over many years. A long-term decline in the water table [Figure 1.8(a)] can result from reduced recharge (e.g., extended drought) or increased dis-

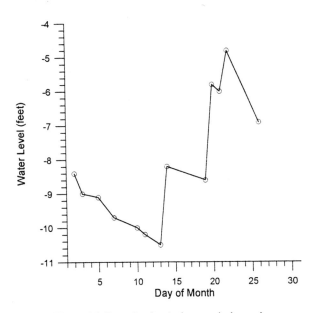

Figure 1.6 Example of a single-storm hydrograph.

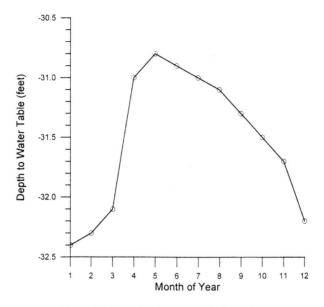

Figure 1.7 Example of a seasonal hydrograph.

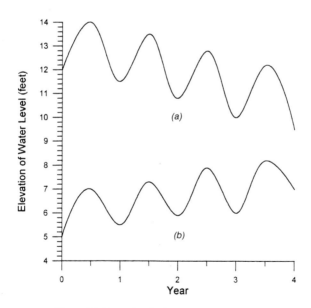

Figure 1.8 Example of a long-term hydrograph.

charge (e.g., over pumpage); whereas a long-term rise [Figure 1.8(b)] can be brought about by increased recharge (e.g., field irrigation). [For additional information, see Driscoll (1986), pp. 69–71 and Heath (1987), pp. 14–15.]

1.5 WATER TABLE GEOMETRY

Figure 1.9 shows that the configuration of the water table conforms roughly to the topography of the land surface; the water table is mounded beneath hills and depressed beneath valleys. Thus, the water table slopes away from drainage divides and toward streams, lakes, and swamps.

The slope of the water table (or the potentiometric surface) in the direction of the steepest change is known as the *hydraulic gradient* (see Figure 1.10). Values of hydraulic gradient are expressed in consistent units, such as feet per foot. Hydraulic gradient is important because it not only defines the direction of groundwater flow but also influences the rate of flow.

A plan view of the configuration of the water table is illustrated by a *water table map*. A water table map is similar in appearance to a topographic map, but the contour lines of a water table map connect points of equal groundwater elevation rather than points of equal land elevation. Figure 1.11 displays an example of a typical water table map. Point A depicts a discharging well which has lowered the regional groundwater table in the immediate vicinity of the well. Point B represents an artificial recharge area in which irrigation water has produced a "mounding" of the regional water table. For confined aquifers, where artesian conditions exist, a contour map of the potentiometric surface may be constructed. The spacing of groundwater contour lines is indicative of both the hydraulic gradient and the permeability of the aquifer (see also Section 1.6). Closely spaced contour lines indicate a steep hydraulic gradient and are likely to be associated with low values of permeability. Widely spaced

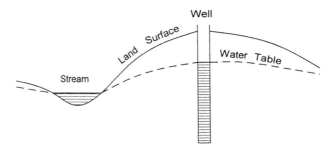

Figure 1.9 General relationship between topography and water table.

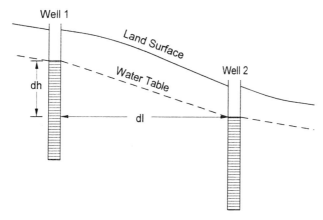

Hydraulic Gradient = dh/dl

Figure 1.10 Illustration of hydraulic gradient.

contour lines indicate a gentle hydraulic gradient and are typically associated with high permeability values. [For additional information, see Heath (1987), pp. 20–23 and 31.]

1.6 ROCK PROPERTIES AFFECTING GROUNDWATER

While solid rock material is the chief object of interest to most field geologists, it is the openings in the rock or soil that are important to hydrogeologists, as these openings are the containers of groundwater. In uncon-

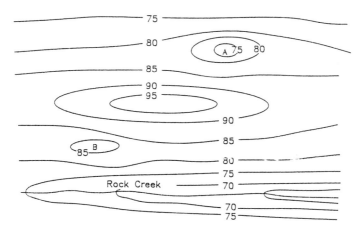

Figure 1.11 Example of a water table map (modified from Davis and DeWiest, 1966, p. 49).

solidated rock (e.g., sand and gravel) openings occur between mineral or rock grains [see Figure 1.12(a–d)] and are called *primary openings*. In consolidated rock (e.g., sandstone, shale, granite, and basalt) the most common openings are fractures [see Figure 1.12(f)] and are called *secondary openings*. Also, limestone may contain large secondary openings in the form of *solution cavities* [see Figure 1.12(e)].

The relative volume of openings in a porous rock material determines the quantity of water the saturated rock can contain. The ratio of openings to the total volume of soil or rock is called *porosity*. Porosity is expressed either in the form of a decimal fraction or a percentage. The following equation yields a value for porosity as a decimal fraction:

$$n = V_v/V_t \qquad (1.1)$$

where n is porosity, V_v is the volume of openings, and V_t is the total volume of the porous material. A percentage value of porosity is obtained by multiplying the decimal fraction by 100.

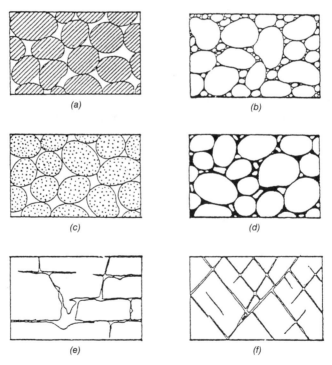

Figure 1.12 Rock texture and groundwater occurrence. (Reprinted from Meinzer, 1942, p. 386 with permission from Dover Publications, Inc.)

TABLE 1.1. Typical Values of Porosity
(from Heath, 1987, p. 7).

Earth Material	Porosity (%)
Soil	55
Clay	50
Sand	25
Gravel	20
Limestone	10
Sandstone	1
Basalt	1
Granite	0.1

Soils and unconsolidated sediments are characterized by intergranular porosity. The intergranular porosity of unconsolidated materials depends on the textural properties of sorting and particle shape, rather than on grain size. Consequently, because fine-grained sediments such as clays and silts tend to be better sorted than coarse-grained materials, they possess greater porosities. The porosity of consolidated rock units is influenced chiefly by the nature and extent of the fractures (e.g., joints, faults, etc.) and is primarily a function of the width of the open fractures. Hence, bedrock possesses fracture porosity or, in the case of karstic limestone, solution porosity. Table 1.1 illustrates some typical values of porosity.

The size and arrangement of openings of the aquifer material influences the ease with which water can pass through it. The measure of the capability of porous rock to transmit water is called the *hydraulic conductivity*. It is defined as the volume of water that will move in unit time under a unit hydraulic gradient through a unit area measured at right angles to the direction of flow (Heath, 1987, p. 12). Its units are those of velocity [i.e., distance divided by time (L/T)].

Values of hydraulic conductivity vary widely from one type of rock to another. For example, highly porous rocks such as karstic (i.e., cavernous) limestones typically exhibit high values of hydraulic conductivity, whereas those of compact shales are low (see Figure 1.13). Even within the same aquifer, widely different values of hydraulic conductivity may be measured. An aquifer in which the hydraulic conductivity is roughly uniform at all locations is known as a *homogeneous aquifer*. An example of a homogeneous aquifer is well-sorted sand. An aquifer that exhibits various values of hydraulic conductivity is called a *heterogeneous aquifer*. Layered sedimentary rock sequences commonly exhibit layered heterogeneity, in which the hydraulic conductivity of the layers is different. Such a sequence is illustrated by an aquifer (high conductivity) bounded above and below by confining units (low conductivity). If the hydraulic conduc-

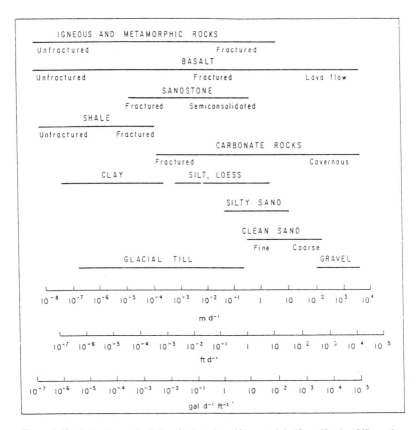

Figure 1.13 Hydraulic conductivity of selected aquifer materials (from Heath, 1987, p. 13).

tivity of an aquifer is uniform in all directions, it is known as an *isotropic aquifer*. Conversely, an *anisotropic aquifer* is one in which hydraulic conductivity varies in different directions. For example, in fractured bedrock hydraulic conductivity is greater in directions parallel to the fractures and lower in directions across the fractures. "Although it is convenient in many mathematical analyses of groundwater flow to assume that aquifers are both homogeneous and isotropic, such aquifers are rare, if they exist at all" (Heath, 1987, p. 13). [For additional information, see Heath (1987), pp. 2–3 and 12–13.]

1.7 DIRECTION OF GROUNDWATER FLOW

The determination of the direction of groundwater flow is an important part of many kinds of hydrogeologic investigations. In particular, it is an essential step in the assessment of groundwater contamination and in the design of groundwater monitoring systems.

Figure 1.14 illustrates a two-dimensional, vertical cross-section of an idealized groundwater flow system. The earth materials of the system are homogeneous and isotropic. The water table is a subdued replica of the topography and slopes downward from a higher position beneath the hills to a lower position beneath the valleys, where it intersects the land surface. This configuration reflects the fact that groundwater flows "downhill" from recharge areas to discharge areas in the direction of the hydraulic gradient. The curved arrows in Figure 1.14 depict typical flow paths.

Like stream basins, groundwater systems are defined by divides, called *groundwater flow divides.* A groundwater flow divide is an imaginary, impermeable boundary across which no flow takes place. At divides situated beneath hilltops or ridgelines, groundwater flow is divergent; whereas, at divides marked by streams or swamps, flow is convergent (see Figure 1.14).

Because groundwater moves from points of high hydraulic head to points of low head (that is, in the direction of the hydraulic gradient), it is a relatively simple matter to determine the direction of groundwater flow through an isotropic aquifer by means of a water level contour map. The contour lines on a water table map (or a potentiometric surface map) connect points of equal hydraulic head and are referred to as *equipotential lines.* The direction of groundwater flow is perpendicular to the equipotential lines in the direction of the hydraulic gradient. This direction may be illustrated by flow lines, which are the imaginary paths that water particles follow. On a water level map, the flow lines cross the equipotential lines at right angles (see Figure 1.15). Additionally, flow lines are oriented perpendicular to the groundwater divides formed by ridgelines and streams.

In the field, both direction of groundwater flow and hydraulic gradient can be readily determined if the following information is known about three wells arranged in a triangular array (see Figure 1.16): the elevation (or total head) at each well, the relative geographic position of each well, and the distance between the wells. The steps of this procedure are described by Heath (1987, p. 11) and in Chapter 8 and are similar to the "Three-Point Problem" well known to structural geologists.

The "grid pattern formed by a network of flow lines and equipotential lines" is referred to as a *flow net* (Heath, 1987, p. 21). A flow net can be used not only to determine the direction of groundwater flow through an aquifer but also to obtain an estimate of the rate of flow (see the following section). Any potentiometric surface map can be developed into a flow net by constructing the equipotential lines so that the decline in head is equal between adjacent pairs of lines, with that flow being equally divided between adjacent pairs of flow lines so as to form a network of squares (see

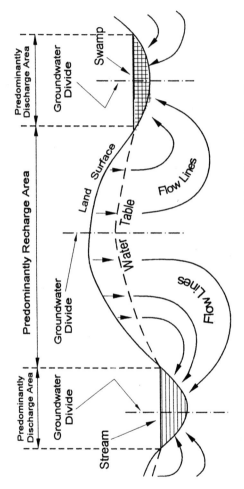

Figure 1.14 Groundwater flow system (modified from Caswell, 1979, p. 12).

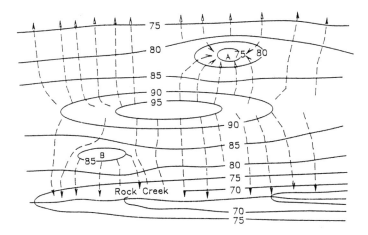

Figure 1.15 Water table map showing groundwater flow lines.

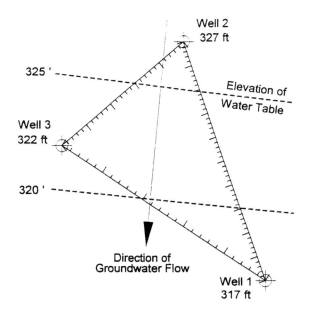

Figure 1.16 Three-point method for determining direction of groundwater flow.

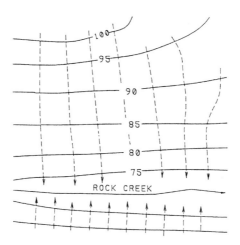

Figure 1.17 Illustration of a simple flow net.

Figure 1.17). While there is an infinite number of equipotential and flow lines in any aquifer, it is necessary to construct only a few of each for the purposes of analyzing groundwater flow. [For additional information, see Heath (1987), pp. 10–11 and 20–23.]

1.8 RATE OF GROUNDWATER FLOW

Determining the rate of groundwater flow is an essential part of numerous kinds of hydrogeologic projects, including water supply investigations and groundwater contamination assessments. For instance, if a contaminant enters an aquifer upgradient of a water supply well, it becomes necessary to estimate the time when the contaminant will reach the well.

Given certain simplifying assumptions, it is possible to estimate the rate of water flow through a specified region of an aquifer by use of Darcy's Law. Darcy's Law states that the rate of flow through a porous medium is directly proportional to the loss of hydraulic head and inversely proportional to the flow length. That is to say, flow rate is proportional to the hydraulic gradient. Darcy's law is typically expressed as follows:

$$Q = KA(dh/dl) \tag{1.2}$$

where Q is the volume of water per unit of time [L^3/T], K is the hydraulic conductivity of the aquifer material [L/T], A is the cross-sectional area, at a right angle to the flow direction [L^2], dh is the difference in hydraulic head between two points [L], *dl* is the distance along the flow path between two points [L], and *dh/dl* is the hydraulic gradient [L/L]. Additionally,

$$A = bW \tag{1.3}$$

where *b* is the saturated thickness of the aquifer [L] and *W* is the width of the aquifer [L]. If distance is measured in feet then hydraulic gradient will be stated in feet/day and the rate of flow will be in cubic feet/day. In the strict sense, Darcy's Law is applicable only when flow is laminar, but this condition poses no serious problem for most natural groundwater flow.

In groundwater contamination problems, it is commonly necessary to estimate how long it will take for a contaminant to move from one point to another. An estimation of the velocity of groundwater flow can be obtained from the equation

$$v = \frac{K(dh/dl)}{n} \tag{1.4}$$

where *v* is the velocity of groundwater flow [L/T] and *n* is the porosity of the material [dimensionless] (other terms have been defined previously). The values of flow velocity that are calculated in this fashion must be considered crude estimates at best. Depending on the density of the contaminant and other factors, a groundwater contaminant may have faster or slower flow rates. The travel time between any two specified points is estimated by multiplying the velocity of flow, *v*, times the distance between the two points, *dl*. [For additional information, see Driscoll (1986), pp. 79–85 and Heath (1987), pp. 10–13.]

1.9 REFERENCES

Caswell, W. B., 1979. *Groundwater Handbook for the State of Maine.* Maine Geological Survey, Augusta, Maine.

Chandler, J. 1990. The hydrological cycle. *Water Well Journal,* January, pp. 44–45.

Davis, S. N., and DeWiest, R. J. M., 1966. *Hydrogeology.* New York, NY: John Wiley & Sons.

Driscoll, F. G., 1986. *Groundwater and Wells* (2nd ed.). St. Paul, MN, Johnson Division.

Freeze, R. A., and Cherry, J. A., 1979. *Groundwater.* Englewood Cliffs, NJ: Prentice-Hall.

Heath, R. C. 1987. *Basic Ground-Water Hydrology.* U.S. Geological Survey Water-Supply Paper 2220.

Lohman, S. W., and others, 1972. *Definitions of Selected Ground-Water Terms — Revisions and Conceptual Refinements.* U.S. Geological Survey Water-Supply Paper 1988.

Meinzer, O. E. (Ed.), 1942. *Hydrology.* New York, NY: Dover Publications.

Todd, D. K., 1980. *Groundwater Hydrology.* New York, NY: John Wiley & Sons.

CHAPTER 2

Case Study: Hydrogeologic Investigation of a Proposed Flyash Storage Site

2.1 STATEMENT OF THE PROBLEM

A large electric-utility company has selected a tract of land adjacent to one of its power plants as a site for the landfill disposal of flyash. Stored in a landfill, flyash will eventually come into contact with infiltrating water and create a leachate, which can pose a contamination threat to groundwater supplies. In order to assess the threat of groundwater contamination and to implement measures to reduce this threat, a thorough knowledge of the hydrogeologic conditions of the storage site becomes essential. This knowledge must include a detailed characterization of groundwater occurrence and movement.

2.2 PURPOSE AND SCOPE

The purpose of this project is to characterize the occurrence and movement of groundwater beneath the site of the proposed flyash storage facility, in order to assess the potential for groundwater contamination. The specific objectives are to

- define the geometry of the water table, including elevation and depth below the land surface
- describe the seasonal fluctuation of the water table
- describe the direction of groundwater flow
- estimate the rate of groundwater flow
- define any regions of the site that do not exhibit a minimum of two feet of vertical separation between the land surface and the seasonal-high water table

[For additional information, see Cherkauer (1980). This paper describes the results of a similar study into the effects of flyash disposal on ground-water.]

2.3 BACKGROUND AND PROJECT SETTING

Flyash is a powdery combustion residue caught by the exhaust gas cleaning equipment of coal burning utility plants. Morphologically, it consists predominantly of silt-sized sub-spherical particles. Chemically, it may be highly variable, but is generally composed of major constituents such as silicon, aluminum, iron, and calcium oxides, and minor amounts of heavy metals such as copper, zinc, nickel, and lead.

The power plant will produce nearly 250,000 tons of flyash a year. The dryflash will be conveyed from the plant to temporary storage silos by a pneumatic system and then trucked to a storage site adjacent to the power plant, where it will be collected in 15-foot high lifts and compacted in order to minimize dust problems and to optimize storage capacity.

The proposed storage site will be engineered to ensure the isolation of the flyash from groundwater (see Figure 2.1). First, the topsoil at the site will be removed; then, the subsoil will be broken up and recompacted into a subbase 2-feet thick and at least 2 feet above the seasonal-high water table. A synthetic liner will be installed above the subbase to prevent leachate from reaching the groundwater. Additionally, a drainage system will be constructed above the liner to collect the leachate and channel it to a collection pond for later treatment. Finally, topsoil and grass seed will be added to the side of the storage pile while the storage site is in use and then to the top after it has beeen closed. During operation and following closure of the site, periodic sampling and analysis of groundwater from

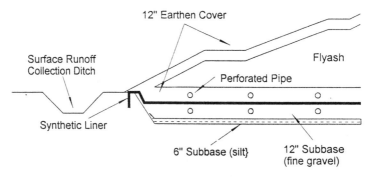

Figure 2.1 Example of a dry flyash disposal system.

monitoring wells will be conducted in order to ensure that the facility is operating properly and the environment is protected.

The proposed storage site is situated on approximately 75 acres of gently rolling agricultural land in the Valley and Ridge Province of the Appalachian Mountains. It is underlain chiefly by moderately fractured, black and dark gray shales of the Marcellus Formation (Devonian). Soil cover is thin — less that 12 inches. A small perennial stream bisects the site along a north-to-south line, and overland flow is generally toward the east and west into this stream. Exploratory examination of the site indicates that the aquifer is unconfined. The climate of the region is temperate, characterized by warm summers, cold winters, light surface winds, and a relatively even distribution of precipitation throughout the year. The average annual precipitation is 39 inches.

2.4 METHODS OF INVESTIGATION

In order to define the configuration of the water table beneath the storage site, to monitor fluctuations in the water table, and to estimate the direction and rate of groundwater flow, the project employs the following methods of investigation.

(1) Installation of twelve shallow piezometers. Figure 2.2 illustrates a topographic base map of the project site, which displays the locations of the piezometers. [The design and construction of monitoring wells are discussed by Aller et al. (1989), Barcelona et al. (1983), and Driscoll (1986).]

(2) Measurement of the depth to water in each piezometer biweekly for a period of approximately one year prior to the start of construction (see Chapter 3).

(3) Calculation of the elevation of the water table (above mean sea level) and the depth to the water table below the land surface from the field measurements of depth to water in the piezometers (see Chapter 4).

(4) Construction of groundwater hydrographs, which illustrate the fluctuation of the water table and permit the identification of the seasonal-high water level (see Chapter 5).

(5) Construction of a water table map, which shows the configuration of the seasonal-high water table (see Chapter 6).

(6) Construction of a depth to water map, which permits the identification of any regions of the site exhibiting less than 2 feet of vertical separation between the land surface and the seasonal-high water table (see Chapter 6).

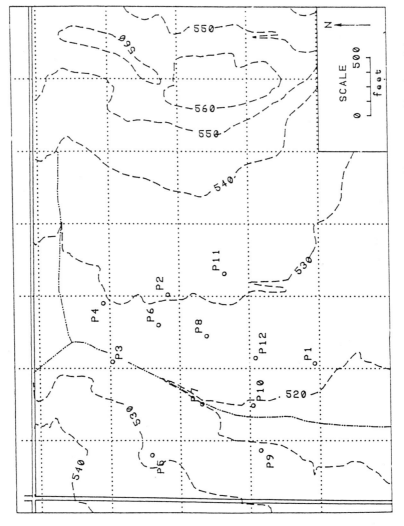

Figure 2.2 A base map of the proposed flyash storage site displaying topographic contours and locations of piezometers.

TABLE 2.1. An Example of an Outline of a Hydrogeologic Report.

Section	Products and Exhibits
TITLE PAGE	
1.0 INTRODUCTION	
1.1 Purpose and Scope	
1.2 Project Setting	Index map of project site
2.0 METHODS OF INVESTIGATION	
2.1 Field Measurements	Map of piezometer locations
2.2 Data Analysis and Presentation	
3.0 RESULTS OF INVESTIGATION	
3.1 Description of Water Table Geometry	Table(s) of water level data
	Groundwater hydrograph(s)
	Water table map(s)
	Depth to water map(s)
3.2 Analysis of Groundwater Flow	Flow line map(s)
	Table(s) of flow rate calculations
4.0 CONCLUSIONS AND RECOMMENDATIONS	
4.1 Summary of Major Findings	
4.2 Recommendations	
5.1 REFERENCES	

(7) Construction of a flow line map, which illustrates the direction of groundwater flow throughout the storage site (see Chapter 7).

(8) Estimation of the rate of groundwater flow at the site (see Chapter 8).

2.5 REPORT OF FINDINGS

Table 2.1 displays an example of a typical format and outline of a report that summarizes the results of a hydrogeologic investigation of the flyash storage site or of a similar project.

2.6 REFERENCES

Aller, L., Bennett, T. W., Hackett, G., Petty, R. J., Lehr, J. H., Sedoris, H., Nielsen, D. M., amd Denne, J. E., 1989. *Handbook of Suggested Practices for the Design and Installation of Ground-Water Monitoring Wells.* Dublin, OH: National Water Well Assoc.

Barcelona, M. J., Gibb, J. P., and Miller, R., 1983. Guide to the Selection of Materials for Monitoring Well Construction and Ground-Water Sampling. Champaign, IL: Illinois State Water Survey.

Cherkauer, D. C., 1980. The effect of flyash disposal on a shallow ground-water system. *Ground Water,* 18(6):544–550.

Driscoll, F. G., 1986. *Groundwater and Wells* (2nd ed.). St. Paul, MN: Johnson Division.

CHAPTER 3

Field Procedure for Measuring Depth to Water Level in Wells and Piezometers

3.1 INTRODUCTION TO METHODS

MEASURING depth to water level in wells or piezometers is the first step necessary in order to

- define the elevation and configuration of the water table or the potentiometric surface
- describe temporal fluctuations of groundwater levels and define seasonal-high water table
- determine the direction of groundwater movement
- estimate the velocity of groundwater flow

For a background discussion of depth to groundwater, see Section 1.3.

This chapter describes two common and simple methods of measuring depth to water level in wells or piezometers: the *Wetted Tape Method* (Section 3.2) and the *Electric Sounder Method* (Section 3.3). The Wetted Tape Method [Figure 3.1(a)] is a modest and inexpensive technique for measuring depth to water in shallow wells in which the approximate depth of the water level is known. The Electric Sounder Method [Figure 3.1(b)] employs a water-level indicator on which a light or buzzer signals a closed circuit when the probe touches water. This method is more rapid and accurate than the wetted tape method, and is more efficient in deep wells or for multiple measurements. The methods described here may not be suitable for contaminated wells. A special interface meter may be necessary in wells where a layer of hydrocarbon product, such as gasoline or oil, may be floating on the water.

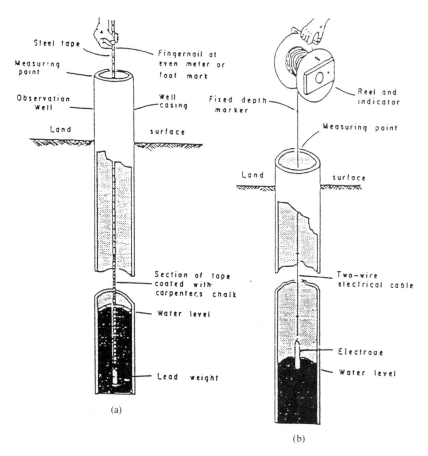

Figure 3.1 Two common methods of measuring depth to water in wells (from Heath, 1987, p. 72).

3.2 WETTED TAPE METHOD

Purpose

To measure the depth to water level in a well or piezometer

Reference

Driscoll, F. G., 1986. *Groundwater and Wells* (2nd ed.). St. Paul, MN: Johnson Division, pp. 549–550.
Heath, R. C., 1987. *Basic Ground-Water Hydrology.* U.S. Geological Survey Water-Supply Paper 2220, pp. 72–73.

Equipment and Materials

- steel surveyor's tape, with weighted end
- carpenter's chalk
- data form (e.g., Figure 3.2)
- pencil (3H)

Procedure

(1) Attach a lead weight to the free end of a steel surveyor's tape.

(2) Smear the free end of the tape with carpenter's chalk over a length of approximately 3 feet.

(3) Lower the tape into the well until the chalked end is immersed in water [see Figure 3.1(a)].

(4) Read the value of the scale division on the tape where it marks the measuring point at the top of the well casing, designated as the reference point [see Figure 3.1(a)].

(5) Withdraw the tape from the well and determine the depth to water level in the well by subtracting the value at the water mark on the chalked end of the tape from the value of the scale division that marked the measuring point. (The demarcation between the wetted and unwetted parts of the tape should be obvious.)

(6) Record this measurement and the well number or identification code on the data form (e.g., Figure 3.2).

(7) Repeat Steps 2–6 in the same well. If the two measurements of the water level made within a few minutes of each other do not agree within ±0.01 feet, continue to make measurements until reliable results are obtained.

(8) Repeat the method for all of the wells.

3.3 ELECTRIC SOUNDER METHOD

Purpose

To measure the depth to water level in a piezometer or well.

Reference

Driscoll, F . G., 1986. *Groundwater and Wells* (2nd ed.). St. Paul, MN: Johnson Division, p. 549.

DEPTH TO WATER MEASUREMENTS
page __ of __

Project: _____

Location: _____

Date: _____

Water levels
measured by: _____

Well Number	Elev. of Top of Casing feet	Height of Casing above Land Surface feet	Depth to Water feet

Figure 3.2 Data form for depth to water measurements.

30

Heath, R. C., 1987. *Basic Ground-Water Hydrology.* U.S. Geological Survey Water-Supply Paper 2220, pp. 72–73.

Equipment and Materials

- water-level indicator (electric sounder)
- carpenter's tape or 12-inch scale
- data form (e.g., Figure 3.2)
- pencil (3H)

Procedure

(1) Turn on the power switch of the water-level indicator and adjust the sensitivity knob, if necessary.

(2) Lower the probe end of the cable into the well until the sounder (i.e., light or buzzer) is activated when the electrode touches the water [see Figure 3.1(b)].

(3) Slowly withdraw the cable until the sounder is deactivated. Then carefully lower the cable again until the sounder signals that the end of the probe is positioned at the top of the water.

(4) Read the value of the scale division on the cable where the cable marks the top of the well casing (i.e., the reference point). This value is a direct measurement of the depth to water in the well. (The cables on some water-level indicators are scaled only at large divisions, such as feet; smaller divisions may have to be measured with a scale laid along the side of the cable.)

(5) Record this measurement and the well number or identification code on the data form (see, e.g., Figure 3.2).

(6) Repeat Steps 2–5 in the same well. If the two measurements of the water level made within a few minutes of each other do not agree within ±0.01 feet, continue to make measurements until reliable results are obtained.

(7) Repeat the procedure for all of the wells.

3.4 REFERENCES

Driscoll, F. G., 1986. *Groundwater and Wells* (2nd ed.). St. Paul, MN: Johnson Division.

Heath, R. C., 1987. *Basic Ground-Water Hydrology.* U.S. Geological Survey Water-Supply Paper 2220.

CHAPTER 4

Analytical Procedure for Determining Elevation of Water Level and Depth to Water Level below Land Surface

4.1 INTRODUCTION TO METHODS

IN order to be useful for the construction of hydrographs and contour maps of the water table, field measurements of depth to water level in wells must be transformed into data expressing elevation of the water level and depth to water level below land surface. For a background discussion of depth to groundwater, see Section 1.3.

This chapter describes two methods by which measurements of depth to water in wells or piezometers may be transformed into elevation and depth below land surface data: the *Electronic Calculator and Worksheet Method* (Section 4.2) and the *Microcomputer and Electronic Spreadsheet Method* (Section 4.3). The Electronic Calculator and Worksheet Method is simple and may be employed readily in the field or under other circumstances when no microcomputer is available. The Microcomputer and Electronic Spreadsheet Method utilizes a special spreadsheet template on which field measurements are entered by the investigator. Calculations of water level elevation and depth of water below land surface are performed automatically by means of formulas stored on the template. By this method a completed data form may be saved on a diskette for later reference and printed out for inclusion in a final project report.

For the application of the methods described in this chapter, use the measurements of depth to water level from the twelve piezometers at the site of the proposed flyash storage facility from the case stude (see Chapter 2), which are displayed in Table 4.1. Alternatively, you may use water level measurements from one of your own projects.

TABLE 4.1. **Depth to Water Level Measurements
for the Flyash Storage Site Case Study.**

Well Number	Elev. of Top of Casing (feet)	Height of Casing above Land Surface (feet)	Depth to Water (feet)
P-1	522.06	2.0	3 ft. 6-1/2 in.
P-2	530.98	0.6	3 ft. 1 in.
P-3	523.99	2.2	1 ft. 10 in.
P-4	529.96	2.7	5 ft. 9 in.
P-5	530.25	2.4	6 ft. 1 in.
P-6	529.61	2.8	6 ft. 8-3/4 in.
P-7	523.88	4.4	5 ft. 1/2 in.
P-8	527.90	2.3	6 ft. 4-3/4 in.
P-9	523.22	2.6	4 ft. 7 in.
P-10	521.81	2.8	4 ft. 5-1/2 in.
P-11	535.10	2.1	4 ft. 7-3/4 in.
P-12	525.70	3.1	5 ft. 8-1/4 in.

4.2 ELECTRONIC CALCULATOR AND WORKSHEET METHOD

Purpose

To calculate elevation of water level, elevation of land surface, and depth to water level below land surface from measurements of the depth to water level in wells or piezometers.

Equipment and Materials

- depth to water level measurements (e.g., Table 4.1)
- electronic calculator
- water level worksheet (e.g., Figure 4.1)
- pencil (3H)

Procedure

(1) Transfer well specifications to the water level worksheet: From a set of depth to water level measurements (e.g., Table 4.1, or use your own field measurements), copy well number, elevation of top casing, and height of casing above land surface for each well to the water level worksheet (Figure 4.1, columns 1, 2, and 3, respectively).

(2) Convert measurement units: If the water level measurements are recorded in units of feet and inches (e.g., 6 ft. 8-3/4 in.), then convert the units to decimal feet (e.g., 6.73 ft.) before entering the value on the

WATER LEVEL CALCULATIONS

Project: _____

Location: _____

Date: _____

Water levels
measured by: _____

1 Well Number	2 Elev. of Top of Casing feet	3 Height of Casing above Land Surface feet	4 Depth to Water in Well feet	5 (2 - 4) Elev. of Water Level feet	6 (2 - 3) Elev. of Land Surface feet	7 (6 - 5) Depth to Water Level below Land Surface feet

Figure 4.1 Water level calculation worksheet.

worksheet in column 4. First, convert fractional inches to decimal inches (e.g., 8-3/4 in. = 8.75 in.). Then, convert decimal inches to decimal feet (e.g., 8.75 in. ÷ 12 in./ft. = 0.73 ft.). Finally, sum the total feet (e.g., 6 ft. + 0.73 ft. = 6.73 ft.).

(3) Copy the measurements of depth to water to the worksheet (column 4).

(4) Determine elevation of water level: Calculate the elevation of water level at each well site by subtracting the depth to water in well (column 4) from the elevation of top of casing (column 2). Record this value in column 5.

(5) Determine elevation of land surface: Calculate the elevation of land surface at each well site, by subtracting the height of casing (column 3) from the elevation of top of casing (column 2). Record this value in column 6.

(6) Determine depth to water below land surface: Calculate the depth to water level below land surface at each well site by subtracting the elevation of water level (column 5) from the elevation of land surface (column 6). Record this value in column 7.

4.3 MICROCOMPUTER AND ELECTRONIC SPREADSHEET METHOD

Purpose

To calculate elevation of the water level and depth to water level below land surface from measurements of the depth to water in wells or piezometers.

Equipment and Materials

- depth to water level measurements (e.g., Table 4.1)
- microcomputer with an 80386 processor or higher, MS-DOS 3.3 or higher, Windows 3.1 or higher, hard drive, floppy disk drive, and graphics printer
- Windows version of electronic spreadsheet software (e.g., Lotus 1-2-3, Quattro Pro, or Excel)
- Hydrodata Diskette

Procedure

This method employs electronic spreadsheet software to calculate elevation of water level and depth to water level below land surface from depth to water level measurements. Field measurements (use Table 4.1 or your own project data) are entered into a spreadsheet template (filename: levtemp.txt), which is loaded from the Hydrodata Diskette. (Instructions for generating diskette are found in the Appendix.) The calculations of elevation of water level and depth to water below land surface are accomplished by means of formulas entered into the cells of the spreadsheet template. When the calculations are complete, a table of the data may be printed.

Starting Up

Turn on the computer, monitor, and printer. Wait until the Windows desktop is displayed.
Open the application.

(1) Open the Windows group icon that contains your software application (e.g., Quattro Pro), or click on the Windows 95 Start button and point first to Programs and then to the application folder.
(2) Double click on the application icon (e.g., Quattro Pro) from the group window, or click on the name of the application from the drop-down list.

Opening the Water Data Template File

(1) Insert the Hydrodata Diskette into drive A (or B).
(2) Choose File | Open. The Open File dialog box will be displayed on the screen.
(3) Click on the arrow next to the File Type list box and select text files (e.g., *.txt) from the drop-down list.
(4) Specify the drive that contains your file by clicking on the arrow next to the Drives list box and then clicking on the drive name (e.g., A).
(5) Specify the file you want to open by selecting the name of the desired file in the file name drop-down box (e.g., levtemp.txt).
(6) Initiate file opening.
For Quattro Pro and Lotus 1-2-3: Click on OK.
For Excel: The Text Import Wizard will appear. Click twice on Next> and once on Finish.

The water level data template will be displayed on the monitor screen. Columns C, D, and E of the template contain the specifications of the twelve piezometers of Table 4.1. For the next part of the exercise, have your depth to water level measurements (from, e.g., Table 4.1) ready to enter into the spreadsheet template. If you wish to enter different well specifications, move the cursor to the appropriate cells in columns D, E, and F and type in the new specifications.

Entering Spreadsheet Formulas

For Microsoft Excel users, be sure to enter an equal sign (=) in front of all formulas given below.

Enter the formula for calculating the elevation of the water level in each piezometer.

(1) Select cell F20.

(2) Type (D20)-(A20 + (B20*0.08333))

(3) Press ENTER.

(4) At cell F20, choose Edit | Copy.

(5) Select the cell range of all piezometers (e.g., F21-F31).

(6) Choose Edit | Paste.

Enter the formula for calculating the depth to water level below land surface in each peizometer.

(1) Select cell G20.

(2) Type (D20-E20)-F20.

(3) Press ENTER.

(4) At cell G20, choose Edit | Copy.

(5) Select cell range of all piezometers (e.g., G21-G31).

(6) Choose Edit | Paste.

Entering Field Measurements

Enter the date on which the water levels were measured.

(1) Select cell C6.

(2) Type the date (if unknown, use today's date).

Enter the name of the person who measured the water levels.

(1) Select cell C8.

(2) Type the name of the field person (e.g., your name).

Enter the number of feet of the first water level measurement.

(1) Select cell A20.

(2) Type the number of feet (e.g., 3).

Enter the number of inches of first water level measurement. If necessary, convert inches to decimal form before entering the values on the worksheet.

(1) Select cell B20.

(2) Type the number of inches (e.g., 6.5).

Enter the number of feet of the second water level measurement.

(1) Select cell A21.

(2) Type the number of feet (e.g., 3).

Enter the number of inches of the second water level measurement.

(1) Select cell B21.

(2) Type the number of inches (e.g., 1.0).

Complete the data entry. Repeat the date entry procedure for each row until the measurements for all piezometers have been typed in. After all the water level data have been entered, compare each of the data entries on the screen with the original measurements and correct any errors.

Modifying Data and Text Format (optional).

Before saving and printing the spreadsheet table, you may modify various properties such as numeric format, column width, text alignment, etc. For example, you may wish to format all data in the columns for two decimal places or center align the column labels. Use the Help menus for specific instructions or refer to the application user's manual.

Saving the Water Level Data File

The water level data should be saved on disk for subsequent retrieval and printing. *Warning!* Do not save the data file under the name of the template file or the original template will be overwritten.

Save the new file for the first time.

(1) Select cell A1.

(2) Choose File | Save As. The Save File dialog box will appear on the screen. The name of an original file (e.g., levtemp.txt) may be displayed and highlighted in the file name text box.

(3) Press BACKSPACE to erase the name of the template file.

(4) Type the new water level data file name (e.g., levdat) in the file name text box.

(5) Click on the arrow next to file type list box and select the type of file used by your particular software program. (Do not use *.txt.).

(6) Click on OK.

The new file will be saved to the drive and directory you specified. Later, to save the existing file choose File | Save. The changes to the file will be saved under the original file name and the old data will be overwritten.

Printing the Water Level Data Form

Before you print, make sure that your application is properly configured for your particular printer. The instructions below provide only basic information about printing the data form. For more detailed instructions, see the application user's manual or click on the Help icon.

Check the printer configuration.

(1) Choose File|Print. The Print dialog box will appear on the screen.

(2) Make sure that your printer is identified as active. (If not, use the printer setup option to select and configure it.)

Edit page settings.

(1) Click on the Page Setup button and, depending on your application, select:

For Quattro Pro: Print Scaling|Print to fit
For 1-2-3 in Size Text box: Fit all to page
For Excel in Scaling Box: Fit to: 1 page

(2) Click on OK.

Preview the print job.

(1) Click on the Print Preview button. A full-page view of the worksheet will be displayed. To view details, use the zoom option.

(2) Press ESC until the screen returns to the worksheet.

Print the data worksheet.

(1) Choose File|Print. The Print dialog box will be displayed. The selected block appears in the Print or Print What box.

(2) Click on Print (or OK).

The Water Level Data Form will be printed on your printer. If the format of the printout requires modification or if an error in data or labels is evident, return to the worksheet and make the necessary corrections. (The application user's guide or reference manual may be useful for these procedures.)

Quitting the Spreadsheet Program

Before closing an application, be sure to save all active files.

(1) Choose File|Exit. If you have saved all active files, the application window closes. If you changed an active file but did not save it, then the Exit (or Save) dialog box will appear.

(2) If the dialog box appears, choose Yes (and, if necessary, Replace) to save the file. The Windows desktop will appear on the screen.

(3) Open another application or quit Windows and turn off the computer. Be sure to close all open applications before quitting Windows.

CHAPTER 5

Method for Graphing Water Level Fluctuations

5.1 INTRODUCTION TO METHODS

HYDROGRAPHS are useful means of visualizing the fluctuations of groundwater levels over a period of time. For a background discussion of water level fluctuations, see Section 1.4.

This chapter describes three methods which can be used to construct groundwater hydrographs: the *Hand Plotting and Graphing Method* (Section 5.2), the *Microcomputer and Electronic Spreadsheet Method* (Section 5.3), and the *Microcomputer and Graphing Software Method* (Section 5.4). The Hand Plotting and Graphing Method is simple but primitive, yet hydrographs of production quality can be created by careful drafting and inking. The Microcomputer and Electronic Spreadsheet Method and the Microcomputer and Graphing Software Method are useful in creating permanent electronic data bases of water levels and provide information for the production of graphs that are suitable for use in project reports or for publication purposes.

Table 5.1 displays seasonal elevations of the water level in Well P-11 of the proposed flyash storage site of the case study (see Chapter 2). These data (or your own water level data) may be used for the application of the methods described in this chapter.

5.2 HAND PLOTTING AND GRAPHING METHOD

Purpose

To construct a hydrograph from groundwater level data.

43

TABLE 5.1. Groundwater Levels for Well P-11
of the Case Study Project.

Date	Time (days)	Elevation (feet)
Feb 14	1	528.77
Feb 28	14	529.10
Mar 18	32	529.19
Apr 01	46	528.35
May 06	81	528.27
May 27	102	528.02
Jun 11	116	527.39
Jun 27	133	527.54
Jul 11	147	527.29
Jul 30	166	527.36
Aug 20	187	526.98
Sep 08	206	526.77
Oct 10	238	527.69
Oct 28	256	528.46
Nov 14	273	529.52
Dec 02	295	529.46
Dec 30	319	529.66
Jan 28	346	529.21
Feb 19	368	528.75

Equipment and Materials

- groundwater level data (e.g., Table 5.1)
- arithmetic graph paper (e.g., Figure 5.1 or K & E 460780)
- 12-inch ruler/scale
- pencil (3H)

Procedure

(1) Orient a sheet of arithmetic graph paper (see, e.g., Figure 5.1) so that the long dimension lies crosswise in front of you.

(2) Label the bottom axis "TIME (days)" and the left axis "ELEVATION (feet)."

(3) Lay out an elevation scale on the left axis of the graph. Establish an axis scale such that the difference between the highest and lowest measured elevations fills as much of the axis as possible, that is, half or more of the axis.

(4) Lay out a time scale on the bottom axis (units = days; do not use calendar dates) and establish scale divisions that fill as much of the axis as possible. (For time periods of approximately one year, use a scale of 1 inch = 50 days.)

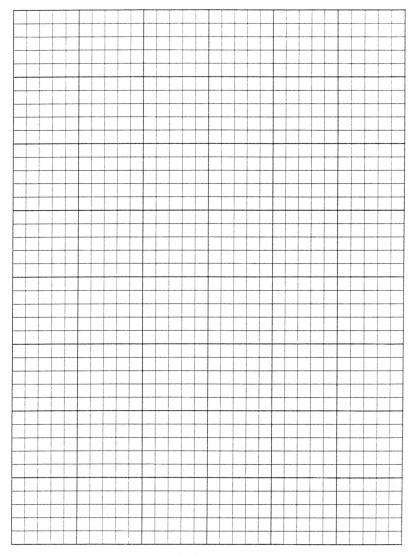

Figure 5.1 Hydrograph form.

(5) Using a list of time-elevation data (see Table 5.1, columns 2 and 3, for example), plot each of the data pairs as a point on the graph.

(6) Complete the hydrograph by connecting all of the data points with a solid line.

5.3 MICROCOMPUTER AND ELECTRONIC SPREADSHEET METHOD

Purpose

To construct a hydrograph from groundwater level data.

Equipment and Materials

- groundwater level data (e.g., Table 5.1)
- microcomputer with an 80386 processor or higher, MS-DOS 3.3 or higher, Windows 3.1 or higher, hard drive, floppy disk drive, and graphics printer
- Windows version of electronic spreadsheet software (e.g., Lotus 1-2-3, Quattro Pro, or Excel)
- Hydrodata Diskette

Procedure

This method employs electronic spreadsheet software to construct a hydrograph from groundwater level data (use Table 5.1 or your own project data). First, a data file of groundwater levels at selected times is created and then a graph of these data is constructed. A copy of the completed hydrograph can be produced on a printer or plotter.

Starting Up

Turn on the computer, monitor, and printer/plotter. Wait until the Windows desktop is displayed.

Open the application.

(1) Open the Windows group icon that contains your software application (e.g., Quattro Pro), or click on the Windows 95 Start button and point first to Programs and then to the application folder.

(2) Double click on the application icon (e.g., Quattro Pro) from the group window, or click on the name of the application from the drop-down list.

(3) Insert the Hydrodata Diskette into drive A (or B).

Entering Groundwater Level Data

The groundwater data are arranged in two columns of data (see Table 5.1, columns 2 and 3, for an example of the data to be entered). The values of time, in days, are entered in column A of the worksheet, and the corresponding values of elevation are entered in column B. Data entry begins with typing in the first time-elevation data pair in the first row of cells (A1 and B1).

Enter the first time value.

(1) Select cell A1.
(2) Type the time value (e.g., 1).

Enter the first elevation value.

(1) Select cell B1.
(2) Type the elevation value (e.g., 528.77).

Enter the second time value.

(1) Select cell A2.
(2) Type the time value (e.g., 14).

Enter the second elevation value.

(1) Select cell B2.
(2) Type the elevation value (e.g., 529.10).

Complete the data entry. Repeat the data entry procedure until all time-elevation pairs have been typed in. After all of the groundwater level data have been entered, compare the data entries on the screen with the original data and correct all errors.

Saving the Groundwater Level Data File

Save the new file for the first time.

(1) Select cell A1.
(2) Choose File|Save As. The Save File dialog box will appear on the screen.
(3) Select the File Name text box and type the new water level data file name (e.g., hydro11).
(4) Click on the arrow next to File Type list box and select the type of file used by your particular software program.
(5) Sepcify the drive that contains the Hydrodata Diskette by clicking on

the arrow next to the Drives list box and then clicking on the drive name (e.g., A).

(6) Click on OK. The new file will be saved to the drive and directory you specified. Later, to save the existing file, choose File|Save.

Creating a Hydrograph

The specific procedures for creating and formatting a hydrograph depend on the particular aplication software that you are using. In the following sections, refer to the instructions which apply to your particular application.

Name the graph/chart and specify the data to be plotted.

Quattro Pro

(1) Choose Graphics|New Graph. The New Graph dialog box will appear.

(2) With Graph Name highlighted, type the name of your graph (e.g., HYDROGRAPH. Do not press ENTER.

(3) Click on the X-Axis arrow. The New Graph dialog box will be minimized and the entire worksheet will be displayed again.

(4) Select (i.e., click and drag) the range of data (e.g., cells A1–A19) that represent the X-axis values.

(5) Press ENTER. The New Graph dialog box will reappear.

(6) Click on the first (series) arrow (or button).

(7) Select the range of data (e.g., cells B1–B19) that represent the first (series) values.

(8) Press ENTER.

(9) If any information for other series is displayed, delete it and click on OK. The worksheet will disappear and a graph window displaying an initial version of the hydrograph will appear on the screen.

(10) Go on to the next section (*Editing the Graph Type*).

1-2-3

(1) Select the range of data (e.g., cells A1–B19) to be graphed, including all headings and labels.

(2) Choose Tools|Chart. The cursor arrow is transformed into a chart pointer.

(3) Click and drag the chart pointer over a blank region of the worksheet (e.g., cells C3–H19). A chart illustrating an initial version of the hydrograph is added to the worksheet. The Chart menu replaces the Range menu on the main menu bar.

(4) Choose Chart | Name.

(5) With the default chart name highlighted, type the name of your chart (e.g., HYDROGRAPH).

(6) Click on Rename.

(7) Go on to the next section (*Editing the Graph Type*).

Excel

(1) Select the range of data (e.g., cells A1–B19) to be graphed, including all headings and labels.

(2) Choose Insert | Chart | On This Sheet. The cursor arrow is transformed into a chart pointer.

(3) Click and drag the chart pointer over a blank region of the worksheet (e.g., cells C3–H19). The ChartWizard dialog box Step 1 (of 5) is displayed.

(4) Choose Next >. The ChartWizard dialog box Step 2 (of 5) is displayed.

(5) Select the XY [Scatter] icon.

(6) Choose Next >. The ChartWizard dialog box Step 3 (of 5) is displayed.

(7) Select format icon #2 [data points and lines].

(8) Choose Next >. The ChartWizard dialog box Step 4 (of 5) and a sample chart is displayed.

(9) Choose Next >.

(10) On the Add a Legend text box, select the No button.

(11) Select the Chart Title text box.

(12) Type a chart title (e.g., HYDROGRAPH – Well P11).

(13) In the Axis Titles text box, select the Category (X): text box.

(14) Type in the title of the X-axis [e.g., TIME (days)].

(15) Select the Value (Y): text box.

(16) Type in the title of the Y-axis [e.g., ELEVATION (feet)].

(17) Choose Finish. An initial version of the hydrograph will be displayed on the worksheet.

(18) Go on the the next section (*Editing the Graph Type*).

Editing the Graph Type

If the displayed graph/chart is not an XY type, then proceed with the instructions in this section.

Quattro Pro

(1) Choose Graphics|Type. The Graph Types dialog box will be displayed.
(2) Select the XY icon.
(3) Choose OK.
(4) Go on to the next section (*Creating Graph Titles*).

1-2-3

(1) Choose Chart|Type.
(2) Select the XY Type button.
(3) Select the upper-left (data points and lines) icon.
(4) Click on OK.
(5) Go on the the next section (*Creating Graph Titles*).

Excel

   The chart type was selected in a previous section. Go on to the next section (*Creating Graph Titles*).

Creating Graph Titles

Quattro Pro

(1) Choose Graphics|Titles (or Graph|Titles). The Graph Titles dialog box will be displayed.
(2) Select the Main Title text box.
(3) Type in the main graph title (e.g., HYDROGRAPH).
(4) Select the Sub Title text box.
(5) Type in the subtitle (e.g., Well P11).
(6) Select X-Axis Title text box.
(7) Type in the X-axis title [e.g., TIME (days)].
(8) Select the Y-Axis Title text box.
(9) Type in the Y-axis title [e.g., ELEVATION (feet)].
(10) Click on OK.
(11) Go on to the next section (*Editing the Graph Format*).

1-2-3

(1) Click anywhere inside the chart frame.
(2) Choose Chart|Headings.
(3) With the Title Line 1 text box highlighted, type in the main chart title (e.g., HYDROGRAPH).

(4) Select the Line 2 text box.

(5) Type in a subtitle (e.g., Well P-11).

(6) Click on OK.

(7) Go on to the next section (*Editing the Graph Format*).

Excel

A chart title was created in a previous section. Go on to the next section (*Editing the Graph Format*).

Editing the Graph Format.

Edit the graph axes.

Quattro pro

(1) Choose Property | X Axis. The X-Axis dialog box will be displayed.

(2) Select Scale.

(3) Select the Increment text box.

(4) Delete all text displayed on the box and type in new data increments (e.g., 40).

(5) Click on OK.

(6) Choose Property | Y Axis. The Y-Axis dialog box will be displayed.

(7) Select Numeric Format.

(8) Select the Fixed button.

(9) Click the arrow next to the Enter Number of Decimal Places text box until the desired number of decimal places (e.g., 1) appears.

(10) Click on OK.

(11) Go on to the next subsection (*Editing the Data Legend*).

1-2-3

(1) Choose Chart | Axis | X-Axis. The X-Axis dialog box will be displayed.

(2) With an initial axis title highlighted, type in the title of the X-axis [e.g., TIME (days)].

(3) In the Scale Manually text box, select the Major Interval text box.

(4) Delete any text displayed in the box and type in new major interval units for the axis (e.g., 40).

(5) In the Scale Manually text box, select the Minor Interval text box.

(6) Delete any text displayed in the box and type in new minor interval units for the axis (e.g., 20).

(7) In the Show Tick Marks text box, select both the Major Interval and Minor Interval boxes.

(8) Click on OK.

(9) Choose Chart | Axis | Y-Axis. The Y-Axis dialog box will be displayed.

(10) With axis title selected (highlighted in blue), type in the name of the Y-axis [e.g., ELEVATION (feet)].

(11) In the Scale Manually text box, select the Major Interval text box.

(12) Delete any text displayed in the box and type in new major interval units for the axis (e.g., 0.5).

(13) In the Scale Manually text box, select the Minor Interval text box.

(14) Delete any text displayed in the box and type in new minor interval units (e.g., 0.1).

(15) In the Show Tick Marks text box, select both the Major Interval and Minor Interval boxes.

(16) Click on OK.

(17) Go on to the next section (*Editing the Data Legend*).

Excel

(1) Double click anywhere inside the chart frame.

(2) Select the X-axis of the graph.

(3) Choose Format | Selected Axis. The Format Axis dialog box will be displayed.

(4) Choose the Patterns tab.

(5) In the Tick Mark Type text box, select the Major Outside button.

(6) Select the Minor Outside button.

(7) Choose the Scale tab.

(8) Select the Major Unit text box.

(9) Delete any text displayed in the box and type in new major time units (e.g., 40).

(10) Select the Minor Unit text box.

(11) Delete any text displayed in the box and type in new minor time units (e.g., 20).

(12) Click on OK.

(13) Select the Y-axis of the graph.

(14) Choose Format | Selected Axis. The Format Axis dialog box will be displayed.

(15) Choose the Patterns tab.

(16) In the Tick Mark Type text box, select Major Outside.

(17) Select Minor Outside.

(18) Choose the Scale tab.

(19) Select the Major Unit text box.

(20) Delete any text displayed in the box and type in new major elevation units (e.g., 0.5).

(21) Select the Minor Unit text box.

(22) Delete any text displayed in the box and type in new minor elevation units (e.g., 0.1).

(23) Click on OK.

(24) Go on to the next section (*Editing the Data Legend*).

Editing the Data Legend

Quattro Pro

No legend is displayed. Go on to the next section (*Editing Grid Lines*).

1-2-3

(1) Choose Chart│Legend, The Legend dialog box will be displayed.

(2) Select the Legend Entry text box.

(3) Delete all text from the box. This action removes a data legend from the chart.

(4) Click on OK.

(5) Go on to the next section (*Editing Grid Lines*).

Excel

No legend is displayed on the chart. Go on to the next section (*Editing Grid Lines*).

Editing Grid Lines

Quattro Pro

(1) Choose Property│X-Axis. The X-Axis dialog box will be displayed.

(2) Select Major Grid Style.

(3) Choose Line Style button.

(4) Select Solid Line Style box.

(5) Click on OK.

(6) Go on to the next section (*Saving the Hydrograph*).

1-2-3

(1) Choose Chart│Grids. The Grids dialog box is displayed.

(2) Click on the arrow to the right of the X-Axis text box.

(3) Select the desired grid-line interval (e.g., Major Interval).

(4) Click on the arrow to the right of the Y-Axis text box.

(5) Select the desired grid-line interval (e.g., Major Interval).

(6) Click on OK.

(7) Go on to the next section (*Saving the Hydrograph*).

Excel

(1) Choose Insert | Gridlines. The Gridlines dialog box will be displayed.

(2) Select the Major Gridlines boxes for both X and Y axes.

(3) Click on OK.

(4) Go on to the next section (*Saving the Hydrograph*).

Saving the Hydrograph and Groundwater Level Data File

Save the graph/chart and data.

(1) Choose File | Save.

The data and the hydrograph will be saved under the original file name (e.g., hydro11).

Printing/Plotting the Hydrograph

Before you print/plot, make sure that your application is properly configured for your particular printer. The instructions below provide only basic information about printing/plotting the hydrograph. For more detailed instructions, see the application reference manual or click on the Help icon.

Check the printer/plotter configuration.

(1) Choose File | Print. The Print dialog box will appear on the screen.

(2) Make sure that your printer or plotter is identified as active. (If not, use the printer setup option to select and configure it.)

Preview the print job.

(1) Click on the Print Preview button. A full-page view of the graph will be displayed. If desirable, modify the appearance of the page by using Page Setup.

(2) Press ESC until the screen returns to the graph.

Print/plot the hydrograph.

(1) Choose File | Print. The Print dialog box will appear on the screen.

(2) Click on Print (or OK). The hydrograph will be produced on your printer or plotter.

Saving a Text (ASCII) File of the Water Level Data

The columns of water level data which were created in this exercise may be useful in other applications, such as a scientific graphing program like GRAPHER. The exchange of data between different applications is facilitated if the data are in text (ASCII) form.

(1) Choose File | Save As. The Save File dialog box is displayed.

(2) Click on the arrow next to the File Types list box and select the name and extension that designates text files.

(3) Select the File Name text box.

(4) If necessary, delete all text in the box and type in a new file name (e.g., hydro11.txt).

(5) Click on OK. A dialog box will appear with a message which warns you that graphs cannot be saved in the ASCII format; that is, only the data of the worksheet will be saved.

(6) Click on OK or Write.

(7) The water level data will be saved as a text (ASCII) file.

Closing the Spreadsheet Application and Quitting Windows

Before closing the application, be sure to save all active files.

(1) Choose File | Exit. If you have saved all active files, the application window closes. If you changed an active file but did not save it, then the Exit dialog box will appear.

(2) If the Exit dialog box appears, choose Yes to save the file. The Windows desktop will appear on the screen.

(3) Open another application or quit Windows and turn off the computer. Be sure to close all applications before quitting Windows.

5.4 MICROCOMPUTER AND GRAPHING SOFTWARE METHOD

Purpose

To construct a hydrograph from data of elevations of the groundwater level in a piezometer or well.

Equipment and Materials

- groundwater level elevation data (e.g., Table 5.1)
- microcomputer with an 80386 processor or higher, MS-DOS 3.3 or higher, Windows 3.1 or higher, hard drive, floppy disk drive, and graphics printer
- GRAPHER for Windows (Version 1.25)
- Hydrodata Diskette

Procedure

This method employs GRAPHER for Windows to construct a groundwater hydrograph of water level elevation data. The GRAPHER worksheet can be used to create and save a data file of groundwater elevations (use Table 5.1 or your own project data). If you are entering water level data from the keyboard, refer to *Entering Groundwater Level Data*. Alternatively, an ASCII data file of groundwater elevations (e.g., hydro11.txt which was created in Section 5.3) may be imported and used to construct a hydrograph. [See *Opening a Text (ASCII) File of Groundwater Level Data* which follows.]

Starting Up

Turn on the computer, monitor, and printer. Wait until the Windows desktop is displayed.

Open the application.

(1) Open the Windows group icon that contains your software application (e.g., Golden Software), or click on the Windows 95 Start button and point first to Programs and then to the application folder.

(2) Double click on the application icon (e.g., GRAPHER) from the group window, or click on the name of the application from the drop-down list. The GRAPHER Plot1 window will be displayed. (If the page frame appears in landscape orientation, you should reorient it to portrait. Choose File|Page Layout. Select Portrait and click on OK.)

(3) Insert the Hydrodata Diskette into drive A (or B).

Display the GRAPHER worksheet.

(1) Choose File|Worksheet. A worksheet window (labeled Sheet1) will appear superimposed over the plot window. If you have a text (ASCII) file of groundwater level data (e.g., filename: hydro11.txt), you may skip the next section of instructions and go on to Opening a Water Level Data File. If you must create a new data file for GRAPHER, then proceed with the following steps.

Entering Groundwater Level Data

The groundwater data are arranged in two columns of data (see Table 5.1, columns 2 and 3, for an example of the data to be entered). The values of time, in days, are entered in column A of the worksheet, and the corresponding values of elevation are entered in column B. Data entry begins with typing in the first time-elevation data pair in the first row of cells (A1 and B1).

Enter the first time value.

(1) Select cell A1.

(2) Type the time value (e.g., 1).

Enter the first elevation value.

(1) Select cell B1.

(2) Type the elevation value (e.g., 528.77).

Enter the second time value.

(1) Select cell A2.

(2) Type the time value (e.g., 14).

Enter the second elevation value.

(1) Select cell B2.

(2) Type the elevation value (e.g., 529.10).

Complete the data entry.

Repeat the data entry procedure until all time-elevation pairs have been typed. After all the groundwater level data have been typed in, compare the data entries on the screen with the original data and correct all errors. When the data file is completed, save it to your data disk.

Saving the Groundwater Level Data File

Save the new file for the first time.

(1) Select cell A1.

(2) Choose File | Save As. The Save As dialog box will appear on the screen.

(3) Press BACKSPACE to delete any displayed text and type the new file name (e.g., hydro11) in the file name text box.

(4) Click on the arrow next to file type list box and select ASCII files [*.DAT].

(5) Click on the arrow next to the Drives list box and select the drive that contains the Hydrodata Diskette (e.g., A or B).

(6) Click on OK. The new file will be saved to the drive and directory you specified. Later, to save the existing file choose File│Save. Skip the next section and go on to *Creating a Hydrograph.*

Opening a Text (ASCII) File of Groundwater Level Data

(1) If you have not inserted the Hydrodata Diskette into drive A (or B), do so now.
(2) Choose File│Open. The Open Data dialog box will be displayed on the screen.
(3) Click on the arrow next to the File Type list box and select ASCII data files (*.txt) from the drop-down list.
(4) Specify the drive that contains your file by clicking on the arrow next to the Drives list box and then double clicking on the drive name (e.g., A).
(5) Specify the drive that contains your file by clicking on the arrow next to the Drives list box and then double clicking on the drive name (e.g., A).
(6) Specify the file you want to open by selecting the name of the desired file in the file name drop-down box (e.g., hydroll.txt).
(7) Click on OK. The water level data will be displayed in columns A and B.

Creating a Hydrograph

(1) Choose Window│Plot1
(2) Choose Graph│Line or Symbol. The Select Worksheet dialog box will be displayed.
(3) Select the Water Level text file (e.g., hydroll.txt).
(4) Click on OK. The Line Plot dialog box will be displayed. This box contains several specifications of the graph. (For more information, refer to the application user's manual.)
(5) Click on OK. The dialog box will disappear and an initial version of the hydrograph will be displayed.

Editing the Graph Format

Edit the X-Axis.
(1) Select (i.e., click on) the X-axis of the graph.
(2) Choose Set│Axis. The Edit X Axis dialog box will be displayed.

(3) In the Length and Starting Position text box, click on the arrow next to the Length text box until it reads 5.00 in.

(4) In the Length and Starting Position text box, click on the arrow next to the X: text box until it reads 2.00 in.

(5) In the Length and Starting Position text box, click on the arrow next to the Y: text box until it reads 3.50 in.

(6) Select the Title text box.

(7) Type in the title of the X-axis [e.g., TIME (days)].

(8) Choose the Edit Ticks button. The Tick Marks dialog box will be displayed.

(9) Select the Spacing (data units) text box in the Major text box.

(10) Delete the default number (e.g., 100) and type in a new spacing value (e.g., 40).

(11) Click on OK.

(12) Choose the Edit Labels button. The Tick Labels dialog box will be displayed.

(13) Choose the Format button. The Label Format dialog box will appear.

(14) Click on the arrow next to the Decimal Digits text box until a zero (0) is displayed.

(15) Click on the OK button of each dialog box until the plot window is displayed again.

Edit the Y-Axis

(1) Select the Y-axis of the graph.

(2) Choose Set|Axis. The Edit Y Axis dialog box will be displayed.

(3) In the Length and Starting Position text box, click on the arrow next to the Length: text box until it reads 4.00 in.

(4) In the Length and Starting Position text box, click on the arrow next to the X: text box until it reads 2.00 in.

(5) In the Length and Starting Position text box, click on the arrow next to the Y: text box until it reads 3.50 in.

(6) Select the Title text box.

(7) Type in the title of the Y-axis [e.g., ELEVATION (feet)].

(8) Choose the Edit Ticks button. The Tick Marks dialog box will be displayed.

(9) Select the Spacing (data units) text box in the Major text box.

(10) Delete the default number (e.g., 2) and type in a new spacing value (e.g., 0.5).

(11) Click on OK.

(12) Choose the Edit Labels button. The Tick Labels dialog box will be displayed.

(13) Choose the Format button. The Label Format dialog box will appear.

(14) Click on the arrow next to the Decimal Digits text box until a one (1) is displayed.

(15) Click on the OK button of each dialog box until the plot window is displayed again.

(16) Select View | Zoom Page. This action will resize the hydrograph so that the entire page is in view.

Edit the Line/Symbols

(1) Select the line that marks the trend of the plotted data.

(2) Choose Set | Symbol Attributes. The Symbol Attributes dialog box will be displayed

(3) Select the icon of the plot symbol you want to use (e.g., solid square).

(4) Click on OK.

Editing Grid Lines

(1) Select the X-axis.

(2) Choose Set | Grid Lines. The Grid Lines dialog box will be displayed.

(3) Select the At Major Ticks box.

(4) Click on OK.

(5) Select the Y-axis.

(6) Choose Set | Grid Lines. The Grid Lines dialog box will be displayed.

(7) Select the At Major Ticks box.

(8) Click on OK.

Adding a Graph Title

(1) Choose Draw | Text. The normal cursor arrow will change to an arrow with a T.

(2) Position the cursor somewhere near the page coordinates $X = 2$, $Y = 9$. (The exact location is not necessary because the graph title may be repositioned at a later time.)

(3) Click on the chosen location. The Text dialog box will be displayed.

(4) Click on the arrow next to the Points text box until the default number (e.g., 12) is replaced by a new font size (e.g., 24).

(5) In the Text box at the bottom of the dialog box, type in the graph title (e.g., HYDROGRAPH—Well P-11) in the space following the blinking cursor bar.

(6) Click on OK.

(7) On the hydrograph, select the graph title.

(8) Click and drag the title to a suitable location on the page (e.g., centered approximately one inch above the top of the graph).

(9) Unselect the graph title by clicking once outside of the plot frame.

Saving the Hydrograph

Save the graph/chart and data.

(1) Choose File | Save As. The Save As dialog box will be displayed.

(2) With the default file name (e.g., plot1.grf) highlighted, type in the new name for the hydrograph (e.g., hydro11.grf).

(3) Click on OK. The data and the hydrograph will be saved under the original file name.

Printing the Hydrograph

Before you print/plot, make sure that your application is properly configured for your particular printer. The instructions below provide only basic information about printing/plotting the hydrograph. For more detailed instructions, see the application reference manual or click on the Help icon.

Check the printer/plotter configuration.

(1) Choose File | Change Printer. The Change Printer dialog box will appear on the screen.

(2) Select your printer from the Printer list box. (Use Setup to specify print settings.)

(3) Click on OK.

Print/plot the hydrograph.

(1) Make sure that the graph (not the worksheet) is displayed and active.

(2) Choose File | Print.

(3) Click on OK.

Closing the Application and Quitting Windows

Before closing the application, be sure to save all active files.

(1) Choose File | Exit. If you have saved all active files, the application window closes. If you have changed an active file but did not save it, then the Exit dialog box will appear.

(2) If the Exit dialog box appears, choose Yes to save the file. The Windows desktop will appear on the screen.

(3) Open another application or quit Windows and turn off computer. Be sure to close all open applications before quitting Windows.

CHAPTER 6

Method for Preparing Water Level Maps

6.1 INTRODUCTION TO METHODS

W ATER level data obtained from piezometers or wells are readily analyzed and evaluated by means of contour maps, including water table maps, potentiometric surface maps, and depth to water maps. Typically, these contour maps are constructed on base maps of a site or region. For a background discussion of these kinds of maps, see Sections 1.3 and 1.5.

A base map is a map that has been previously prepared on which one can plot pertinent information about a study or project site (land or cultural features, sampling sites, monitoring wells, etc.) and construct various kinds of overlays (distribution of soil and rock units, contour lines, etc.). Figure 6.1 illustrates an example of a base map of a project site, which includes topographic contours, a legend, a north arrow, and graphic scale.

The fundamental data for base maps and water level maps are known as *control points*. Control points are any map positions that can be represented by areal (x and y) coordinates and, commonly, elevation (z) coordinates. These elements include topographic survey stations, monitoring wells, piezometers, geophysical stations, etc. The specific coordinates of control points are established by superimposing a grid on the base map.

This chapter describes two methods by which water level maps may be constructed: the *Hand Drafting Method* (Section 6.2) and the *Microcomputer and Grid-Based Contouring Software Method* (Section 6.3). The Hand Drafting Method employs basic engineering drawing techniques to construct a base map of a project site, on which water level contour lines are then drafted. The Microcomputer and Grid-Based Contouring Software Method uses a gridding, contouring, and three-dimensional surface

63

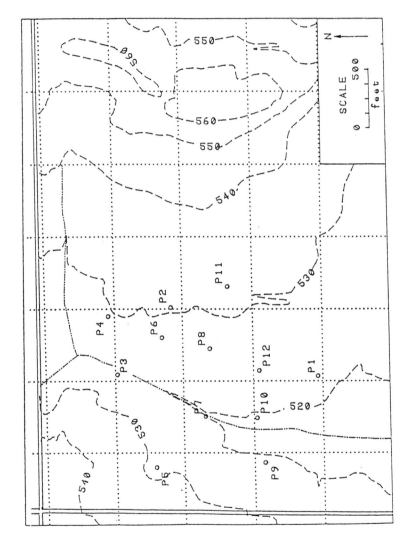

Figure 6.1 A base map of the proposed flyash storage site displaying topographic contours and locations of piezometers.

TABLE 6.1. Control Point Data for the Piezometers of the Case Study.

X-Coordinate	Y-Coordinate	Elevation of Water Level (ft)	Depth to Water Below Land Surface (ft)	Elevation of Land Surface (ft)	Well Number
1064	556	518.52	1.54	520.1	P-1
1565	1631	527.90	2.48	529.4	P-2
1080	2053	522.16	−0.37	521.8	P-3
1500	2115	524.21	3.05	527.3	P-4
407	1756	524.17	3.68	527.7	P-5
1344	1705	522.88	3.93	526.9	P-6
780	1384	518.84	0.64	519.6	P-7
1267	1348	521.50	4.10	525.6	P-8
440	950	518.64	1.98	520.7	P-9
767	1003	517.35	1.66	519.1	P-10
1713	1203	529.45	3.55	533.2	P-11
1106	980	520.01	2.59	522.7	P-12

plotting software application to construct and plot production quality water level maps.

Table 6.1 displays the control point data for a water level study of the flyash storage facility (see the case study in Chapter 2). These data (or your own project data and site) may be used for the application of the methods described in this chapter.

6.2 HAND DRAFTING METHOD

Purpose

To construct water level maps of a project site.

Reference

Dennison, J. M., 1968. *Analysis of Geologic Structures*. New York, NY: W. W. Norton & Co., pp. 68–86.

Equipment and Materials

- map or aerial photograph of a project site (e.g., Figure 6.1)
- grid transparency
- grid or section paper
- control point worksheet (e.g., Figure 6.2)

Name of	Map Coordinates		
Control Point	X	Y	Z

Figure 6.2 Control point worksheet.

- drafting board
- straight-edge ruler
- marking pen (water soluble ink)
- pencil (4H)
- colored pencils (optional)

Procedure

In this method a base map is prepared first from a map or plan of the project site. Then, water level contour maps of the site are constructed either directly on the base map or on overlays. The grid coordinates of control points may be read from a map or air photo of the project site by using a transparency grid.

Establishing the Coordinates of the Control Points

Orient the grid. Align a map (or aerial photograph) of the study site on a drafting board so that the north edge lies at the top and the east edge lies to the right. Fasten the corners of the map or photo securely to the board with short strips of drafting tape.

Select a grid reference point. On the site map locate a permanent marker, such as a benchmark, survey station, etc., at or near the southwest corner of the site (e.g., for the site map on Figure 6.1, use the lower left corner of the map border). Establish this point as the origin of a site grid, with grid coordinates of 0,0.

Specify a map scale and grid spacing. Check the scale of the site map to be sure that it appears in the form 1 in. = some even number of feet (e.g., 1 in. = 200 ft., 1 in. = 400 ft., 1 in. = 2000 ft., etc.). The procedure that follows assumes that the scale is in this form. Then, beginning at the lower left corner of the transparency grid, mark off with a marking pen each 1-inch interval along the bottom and left borders and label each tic mark with the appropriate scale value in feet (e.g., 200, 400, 800, 1200, etc.).

Determine the grid coordinates of selected control points on the site map or plan. Select the features of the site that are to be plotted as control points on the base map. Enter the names of these features in column 1 of the control point worksheet (Figure 6.2). Using the grid transparency determine the x and y coordinates of these control points and record their values on the appropriate lines on the worksheet.

Constructing a Base Map

Plot the reference point of the site on the base map grid paper. Align the base map grid paper on a drafting board so that the north edge lies at the top and the east edge lies to the right. Fasten it securely to the board with short strips of drafting tape. Select the lower left corner of the base map grid paper to represent the reference point of the base map, with grid coordinates of 0,0. Mark this point with a pencil and label the point with its coordinates. You may also mark off and label the coordinates of grid points at 1-inch intervals along the axes of the grid.

Plot control points on the base map. Using the coordinates of the control points from the worksheet (Figure 6.2), locate the position of each control point and mark it with an appropriate symbol (e.g., asterisk, plus sign, etc.). Also, label each of the control points with an identification code (e.g., P-1, P-2, etc.).

Add text and graphics to base map. Complete the base map grid by adding a scale (e.g., 1 in. = 400 ft.), legend or title (e.g., FLYASH STORAGE AREA), north arrow, and other text or graphics, as necessary or desired.

Make additional copies of the base map (optional). If the base map is to be included in a project report, a production quality reproduction can be turned out by making a pencil tracing of the base map on a sheet of vellum or other drafting medium and then inking the drawing. One of more copies of the master map may be readily reproduced on report quality paper on a photocopier and then used for constructing water table and depth to water maps.

Constructing a Water Table Map

Plot the water table elevations on the base map. On a copy of the base map, plot in light pencil the value of the elevation of the water level at each of the control points next to its corresponding symbol.

Construct water table contour lines. Choose a contour interval appropriate to the range of water level elevations. (For example, use a contour interval of 1 foot for elevation ranges of 20 feet or less or a contour interval of 2 feet for ranges of 11–20 feet.) Using a light pencil begin drawing the first contour line near the edge of the map at the lowest elevations and proceed to progressively higher elevations. For simplicity, assume that the difference in elevation between two control points is proportional to the map distance between them. Label each contour line with its elevation value.

Complete the map to report quality. Complete the map by inking in all lines and text.

Constructing a Depth to Water Map

Plot depth to water values on the base map. On a copy of the base map, plot in light pencil the elevation of the depth to water level below land surface at each of the control points next to its corresponding symbol.

Construct depth to water contour lines. Choose a contour interval appropriate to the range of water level depths. (For example, use a contour interval of 1 foot for depth ranges of 10 feet or less or a contour interval of 2 feet for ranges of 11–20 feet.) Begin contouring by drawing the contour line for the maximum value of depth to water and proceed with the contour lines of successively lower depth values. For simplicity, assume that the difference in depth to water between two control points is proportional to the map distance between them. Lable each contour line with its depth value.

Complete the map to report quality. Complete the map by inking in all lines and text. Erase any extraneous pencil marks.

6.3 MICROCOMPUTER AND GRID-BASED CONTOURING SOFTWARE METHOD

Purpose

To construct water level maps of a project site.

Equipment and Materials

- control point data (e.g., Table 6.1)
- microcomputer with an 80386 processor or higher, MS-DOS 3.3 or higher, Windows 3.1 or higher, hard drive, floppy disk drive, and graphics printer
- SURFER for Windows (Version 5.0)
- Hydrodata Diskette

Procedure

This method employs SURFER for Windows to construct water table and depth to water maps from data derived from field measurements. Additionally, topographic contour maps and site base maps can be constructed by using this method. The application creates XYZ data files (use Table 6.1 or your own project data) which are used to construct two-dimensional contour maps and three-dimensional surface maps.

Starting Up

Turn on the computer, monitor and printer/plotter. Wait until the Windows desktop is displayed.

Open the application.

(1) Open the Windows group icon that contains your software application (e.g., Golden Software), or click on the Windows 95 Start button and point first to Programs and then to the application folder.

(2) Double click on the application icon from the group window (e.g., SURFER), or click on the name of the application from the drop-down list.

Creating an XYZ Data File

Open a new worksheet.

(1) Insert the Hydrodata Diskette into drive A (or B).

(2) Choose File|Worksheet. The SURFER-[Sheetl] window will be displayed, replacing the original Plot window. The SURFER data file will consist of XYZ data entered on the application worksheet in six columns. Column A contains the X (west-east) map coordinates of the control points. Column B contains the Y (north-south) map coordinates of the control points. Columns C, D, and E contain the Z values of water level elevation, depth to water, and land surface elevation respectively. Column F contains the control point labels (e.g., P-l).

Enter the column labels.

(1) Select cell A1.

(2) Type in the label of the X data column (e.g., E-W).

(3) Select cell B1.

(4) Type in the label of the Y data column (e.g., S-N).

(5) Select cell C1.

(6) Type in the label of the Z1 data column (e.g., WATRLEV).

(7) Select cell D1.

(8) Type in the label of the Z2 data column (e.g., WATRDEP).

(9) Select cell E1.

(10) Type in the label of the Z3 data column (e.g., TOPOMAP).

(11) Select cell F1.

(12) Type in the label of the control point column (e.g., WELL).

Enter the first X-coordinate value.

(1) Select cell A2.

(2) Type the X-coordinate value (e.g., l064).

Enter the first Y-coordinate value.

(1) Select cell B2.

(2) Type the Y-coordinate (e.g., 556).

Enter the first water level elevation value.

(1) Select cell C2.

(2) Type the water level elevation value (e.g., 518.52).

Enter the first depth to water value.

(1) Select cell D2.

(2) Type the depth to water (e.g., 1.54).

Enter the first land surface elevation value.

(1) Select cell E2.

(2) Type the land surface elevation (e.g., 520.1).

Enter the first well notation.

(1) Select cell F2.

(2) Type the well number (e.g., P-1).

Complete the date entry. Repeat the data entry procedure until all XYZ data have been typed. After all the XYZ values have been typed in, compare the data entries on the screen with the original data and correct all errors. When the data file is completed, save it to your data disk.

Saving the XYZ Data File

Save the new file for the first time.

(1) Select cell A1.

(2) Choose File | Save As. The Save As dialog box will appear on the screen.

(3) Press BACKSPACE to delete any displayed text and type in the new file name (e.g., ashsite).

(4) Click on the arrow next to the file type list box and the type of file used by your application [e.g., ASCII files (*.DAT)].

(5) Click on the arrow next to the Drives list box and select the drive that contains the Hydrodata Diskette (e.g., A or B).

(6) Click on OK. The new file will be saved to the drive and directory you specified. Later, to save the existing file choose File | Save.

Creating a Grid File

Open the data file.

(1) Choose Window | Plotl | Grid | Data. The Open Data dialog box will be displayed.

(2) Specify the drive that contains your XYZ data file by clicking on the arrow next to the Drives list box and then clicking on the drive name (e.g., A).

(3) Select the file you want to open by clicking on the name of the desired file in the file name drop-down box (e.g., ashsite.dat).

(4) Click on OK. The Scattered Data Interpolation dialog box will be displayed. This box designates the gridding parameters to be used by SURFER.

Select XYZ data columns. The X and Y values in columns A and B, respectively, in the Data Columns text box remain fixed for all types of maps; do not change these settings. The choice of the column that contains the Z values depends of the type of map you are constructing. The options are

Water table map	Z = Column C
Depth to water map	Z = Column D
Topographic map	Z = Column E
Base map	Z = Column F

(5) Click on the arrow next to the Z: text box.

(6) Select the appropriate column for the type of map you are constructing (e.g., Column C: WATRLEV).

Specify the grid line geometry. The default values displayed in the Grid Line Geometry text box will create a grid frame (and, consequently, a map) which is only as large as the area occupied by the control points and is most likely to have asymmetrical dimensions. You should modify the grid dimensions to expand the area of the final map; follow the instructions presented below. (If you have no reason to create a map which illustrates an area larger than that occupied by the cluster of control points or which is larger in one areal dimension than the other, then skip the remainder of this subsection and go on to *Name the Grid File*).

(1) Select the X Direction Minimum text box.

(2) Delete the default number and type in a new X dimension minimum value (e.g., 0).

(3) Select the Y Direction Minimum text box.

(4) Delete the default number and type in a new Y dimension minimum value (e.g., 0).

(5) Select the X Direction Maximum text box.

(6) Delete the default number and type in a new X dimension maximum value (e.g., 2400).

(7) Select the Y Direction Maximum text box.

(8) Delete the default number and type in a new Y dimension maximum value (e.g., 2400).

(9) Select the X Direction Spacing text box.

(10) Delete the default number and type in a new X direction spacing value (e.g., 50).

(11) Select the Y Direction Spacing text box.

(12) Delete the default number and type in a new Y direction spacing value (e.g., 50).

Name the grid file.

(1) In the Output Grid File text box (at the bottom of the dialog box), click on the Change button. The Save Grid dialog box will be displayed.

(2) With the default file name (e.g., ashsite.grd) highlighted, press BACK-SPACE to delete all displayed text and type in a new file name; see sample options below.

Water table map:	watrlev.grd
Depth to water map:	watrdep.grd
Topographic map:	topomap.grd
Base map:	basemap.grd

(3) Click on OK. The Scattered Data Interpolation dialog box will return to the screen.

(4) Click on OK. Gridding will begin and the Status dialog box will be displayed. A status bar indicates the progress of the gridding procedure. When the gridding is completed, the Status box will disappear and three beeps will sound.

Creating a Contour Map

(1) Choose Map|Contour. The Open Grid dialog box will be displayed. The grid file created in the previous section of this method (e.g., watrlev.grd) should be displayed in the file name drop-down box.

(2) Click on OK. The Contour Map dialog box will be displayed. Using the default parameters in the Contour Map dialog box will produce an initial version of a contour map.

(3) Click on OK. The dialog box will disappear and the contour map will be drawn in the plot window.

(4) Choose View | Fit to Window.

Editing a Contour Map

Specify contour labels.

(1) Select (i.e., double click on) the contour map. The Contour Map dialog box will be displayed.

(2) In the Contour Levels text box, choose the Label button.

(3) Click on the arrow next to the First Label Contour Line text box until the position of the first contour line to be labeled appears (e.g., 0).

(4) Click on the arrow next to the Labeled Line Frequency text box until the desired labeled line spacing value appears (e.g., 2).

(5) Click on the Format button. The Labels Format dialog box will be displayed.

(6) Click on the arrow next to the Decimal Digits text box until the desired decimal places appear (e.g., 0).

(7) Click on OK in all dialog boxes until the Plot window reappears. A revised contour map will be drawn in the Plot window.

Edit the map axes.

(1) Select (double click on) the bottom axis of the graph. The Bottom Axis dialog box will be displayed.

(2) In the Labels text box, click on the Label Format button.

(3) Click on the arrow next to the Decimal Digits text box until the desired decimal places appear (e.g., 0).

(4) Click on OK in all dialog boxes until the Plot window reappears.

(5) Select the left axis of the graph. The Left Axis dialog box will be displayed.

(6) In the Labels text box, click on the Label Format button.

(7) Click on the arrow next to the Decimal Digits text box until the desired decimal places appear (e.g., 0).

(8) Click on OK in all dialog boxes until the Plot window reappears.

Creating a Scale Bar

(1) Choose Map | Scale Bar.

(2) Select the Cycle Spacing text box.

(3) Delete the default text and type in the desired scale bar intervals (e.g., 400).

(4) Select the Label Increment text box.

(5) Delete the default text and type in the label intervals for the scale bar (e.g., 400).

(6) Click on the Format button.

(7) Click on the arrow next to the Decimal Digits text box until the desired decimal places appear (e.g., 0).

(8) Click on OK.

(9) Choose View | Page.

Creating a Map Title

(1) It may be necessary to shift the plot page toward the top by clicking on the arrow next to the window pane.

(2) Choose Draw | Text. Move the cursor arrow down to the plot page. A T is attached to the arrow.

(3) Position the cursor close by the coordinates X = 3, Y = 9.5 and click on the location. The Text dialog box will be displayed.

(4) Click on the arrow next to the Points text box until the desired font size appears (e.g., 24).

(5) In the text box at the bottom, type the title of the map (e.g., WATER TABLE MAP).

(6) Click on OK. The Plot window will reappear and the map title will be drawn on the page.

(7) Click and drag the Map Title Outline text box until it is centered approximately 1 inch above the center of the top axis of the map.

(8) Click anywhere outside of the map page.

Placing Control Point Symbols on the Map

(1) Click on the map.

(2) Choose Map | Post. The Open Data dialog box will be displayed.

(3) Select the data file that contains the control point (e.g., well or piezometer) labels (e.g., ashsite.dat).

(4) Click on OK. The Post Map dialog box will be displayed.

(5) In the Worksheet Columns text box, click on the arrow next to the Label text box.

(6) Select the column that contains the control point labels (e.g., Column F: WELL).

(7) Click on the Default Symbol button.

(8) Select the appropriate symbol for your control points (e.g., cross-hairs circle).

(9) Click on OK.

(10) In the Symbol Size text box, click on the arrow next to the Fixed Size text box until the desired symbol size appears (e.g., 0.15 in.).

(11) Click on OK. The post map will be drawn as an overlay on the contour map.

(12) Press F2 to select the maps.

(13) Choose Map|Overlay Maps. The two maps will be merged in the proper proportions and the control points (i.e., wells) will be posted on the contour map in their correct positions.

(14) Choose View|Fit to Window.

Saving the Map

Save the map and data.

(1) Choose File|Save As. The Save As dialog box will appear on the screen.

(2) Click on the arrow next to the Drives list box.

(3) With the default file name highlighted, press BACKSPACE to delete any displayed text and type in the new file name (e.g., watrlev.srf).

(4) Click on OK. The new map file will be saved to the drive and directory you specified. Later, to save the existing file choose File|Save.

Printing the Map

Before you print, make sure that your application is configured for the correct printer. The instructions below provide only basic information about printing the map. For more detailed instructions about printing with your particular application, click on the Help icon or refer to your user's manual.

Check the printer/plotter configuration.

(1) Choose File|Print Setup. The Print Setup dialog box will appear on the screen.

(2) Select the printer to use from the Printer list box. (Use Setup to specify print settings.)

(3) Click on OK.

Print/plot the map.

(1) Make sure that the map (not the worksheet) is active.

(2) Choose File | Print. The Print dialog box will appear on the screen.

(3) Select the Truncate button in the Printing method text box.

(4) Click on OK.

Creating a Surface Map (optional)

Construct the map

(1) Choose File | New. The New Window dialog box will be displayed.

(2) Select the Plot button.

(3) Click on OK. A new blank plot window will be displayed.

(4) Choose Map | Surface. The Open Grid dialog box will be displayed.

(5) Select the grid (e.g., watrlev.grd).

(6) Click on OK. The Surface Plot dialog box will be displayed.

(7) In the Base text box, select Show Base and Show Vertical Lines.

(8) Click on OK. The surface map will be drawn in the Plot window.

Edit the axes.

(1) Double click on the bottom (i.e., lower right) axis. (The origin point is positioned at the bottom center of the figure.) The Bottom Axis dialog box will appear.

(2) Click on the Label Format button.

(3) Click on the arrow next to the Decimal Digits text box until the desired decimal places appear (e.g., 0).

(4) Click on OK in all dialog boxes until the Plot page reappears.

(5) Double click on the left axis. The Left Axis dialog box will appear.

(6) Click on the Label Format button.

(7) Click on the arrow next to the Decimal Digits text box until the desired decimal places appear (e.g., 0).

(8) Click on OK in all dialog boxes until the Plot page reappears.

(9) Double click on the Z axis (i.e., left vertical axis). The Z Axis dialog box will appear.

(10) Click on the Label Format button.

(11) Click on the arrow next to the Decimal Digits text box until the desired decimal places appear (e.g., 0).

(12) Click on OK in all dialog boxes until the Plot page reappears.

Scale the map.

(1) Choose Map | Scale.

(2) In the X Scale text box, click on the arrow next to the Length text box

until the new axis length appears (e.g., 5.00 in.). The length of the Y axis changes automatically.

(3) Click on OK. The map will be redrawn at the new scale in the Plot window.

Add a map title.

(1) Choose Draw | Text. Move the cursor arrow down to the plot page. A T is attached to the arrow.

(2) Position the cursor close by the coordinates X = 2.5, Y = 9.5 and click on the location. The Text dialog box will be displayed.

(3) Click on the arrow next to the Points text box until the desired font size appears (e.g., 24).

(4) In the text box at the bottom, type the title of the map (e.g., WATER TABLE SURFACE MAP).

(5) Click on OK. The Plot window will reappear and the map title will be drawn on the page.

(6) Click and drag the Map Title Outline text box until it is centered approximately 1 inch above the center of the top axis of the map.

(7) Click anywhere outside of the map page.

Saving the Map

Save the map and data.

(1) Choose File | Save As. The Save As dialog box will appear on the screen.

(2) With the default file name highlighted, press BACKSPACE to delete any displayed text and type in the new file name (e.g., watr3d.srf).

(3) Click on OK. The new file will be saved to the drive and directory you specified. Later, to save the existing file choose File | Save. The changes to the file will be saved under the original file name and the old data will be overwritten.

Printing the Map

Before you print, make sure that your application is configured for the correct printer. The instructions below provide only basic information about printing the map. For more detailed instructions about printing with your particular application, click on the Help icon or refer to your user's manual.

Check the printer/plotter configuration.

(1) Choose File | Change Printer. The Change Printer dialog box will appear on the screen.

(2) Select the printer to use from the Printer list box. (If necessary, use Setup to specify print settings.)

(3) Click on OK.

Print/plot the map.

(1) Make sure that the map (not the worksheet) is active.

(2) Choose File|Print. The Print dialog box will appear on the screen.

(3) Select the Truncate button in the Printing Method Text box.

(4) Click on OK.

Creating Another Map (optional)

(1) Close all active SURFER files, then choose File|New.

(2) Select the Plot button and click on OK.

(3) To create a new grid and map from an existing data file (e.g., ashsite.dat), go back in the instructions to *Creating a Grid File.* To create a new data file, go back to *Creating an XYZ Data File.*

Closing the Application and Quitting Windows

Before closing the application, be sure to save all active files.

(1) Choose File|Exit. If you have saved all active files, the application window closes. If you changed an active file but did not save it, then the Exit dialog box will appear.

(2) If the Exit dialog box appears, choose Yes to save the file. The Windows desktop will appear on the screen.

(3) Open another application or quit Windows and turn off the computer. Be sure to close all open applications before quitting Windows.

6.4 REFERENCE

Dennison, J. M., 1968. *Analysis of Geologic Structures.* New York, NY: W. W. Norton & Co.

CHAPTER 7

Method for Preparing Flow Line Maps

7.1 INTRODUCTION TO METHODS

T HE direction of groundwater flow is readily illustrated by flow lines, which depict flow direction by a set of arrows. For a background discussion of groundwater flow direction, see Section 1.7.

This chapter describes two methods for constructing groundwater flow line maps on previously prepared base maps: the *Contour Map Method* (Section 7.2) and the *Three Point Method* (Section 7.3). The Contour Map Method employs a water table base map on which groundwater flow lines are drafted by hand. The Three Point Method uses a base map on which the locations of piezometers (or monitoring wells) and the elevation of the water level in each of the piezometers are plotted.

Figure 7.1 illustrates a water table map of the flyash storage facility from the case study in Chapter 2. This map (or a similar map of your own project site) may be used for the application of the Contour Map Method. Figure 7.2 illustrates a control point base map of the flyash storage site. This map (or your own similar base map) may be used for application of the Three Point Method.

7.2 CONTOUR MAP METHOD

Purpose

To construct groundwater flow lines on a water table base map.

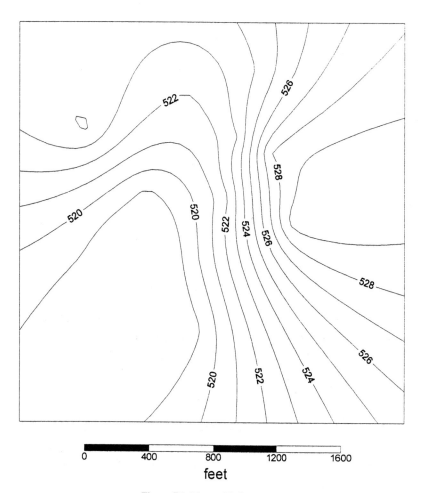

Figure 7.1 Water table base map.

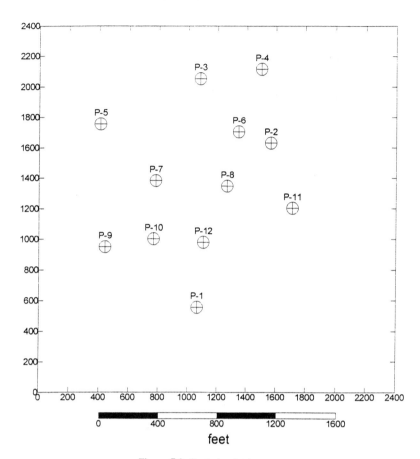

Figure 7.2 Control point base map.

Equipment and Materials

- water table contour map (see, e.g., Figure 7.1)
- pencil (3H)

Procedure

(1) On a water table map (e.g., Figure 7.1) locate the region of the project site that contains the highest contour lines (i.e., region of highest hydraulic head).

(2) Beginning in this region, draw a line to an adjacent region of lower hydraulic head, always keeping the line perpendicular to the contour lines. Typically, the line stretches from a recharge area to a discharge

area, such as a stream, lake, or swamp (see Figure 1.15 for an example).

(3) Complete the drawing of the flow line by adding a barb, which indicates the direction of groundwater flow, on the end of the flow line that is situated at the region of lowest hydraulic head (i.e., in a discharge area).

(4) Repeat this procedure several times throughout the map region until all major groundwater flow directions are illustrated. If possible, space the flow lines so that the areas formed by the intersection of two adjacent flow lines and two adjacent water table contour lines approximate squares.

7.3 THREE POINT METHOD

Purpose

To construct flow lines based on the data from elevations of the groundwater level in piezometers or wells.

Reference

Heath, R. C., 1987. *Basic Ground-Water Hydrology.* U.S. Geological Survey Water-Supply Paper 2220, pp. 72–73.

Equipment and Materials

- control point data of groundwater elevations (e.g., Table 6.1)
- control point base map (e.g., Figure 7.2)
- pencil (3H)

Procedure

By means of the three point method (see Figure 1.16), an estimate of the groundwater flow direction within a small (or uniform) region can be determined with only three groundwater elevations known from wells. The construction requires that the differences in water level elevation and the distances between the three control points be known. With the groundwater elevations known in numerous wells (e.g., Table 6.1), the directions of groundwater flow within larger regions can be estimated by applying the three point method to all wells throughout the region.

(1) Prepare a base map of the project site that displays the location of wells and the elevation of the water table at the locations of the wells (see Chapter 6 for instructions) or use the control point map of the flyash storage site from the case study in Chapter 2 (Figure 7.2).

(2) On the control point base map, label each control point (i.e., well or piezometer) with the value of the water table elevation at that location (e.g., see Table 6.1 for groundwater elevations).

(3) Construct straight lines connecting three control points (e.g., Figure 7.2, control points 522, 524, and 523 in the upper left corner of the map) so that a triangle is formed (see Figure 1.16 for an example).

(4) Subdivide (and mark off with tics) each of the lines in a number of segments equal to the difference in elevation between the control points (see Figure 1.16).

(5) Label each line segment tic mark with its corresponding elevation.

(6) Construct straight lines through the tics of equal elevation (the lines may be dashed as in Figure 1.16). These lines represent lines of equal hydraulic head and correspond to contour lines on a water table map.

(7) The estimated direction of groundwater flow in the vicinity of the three wells is determined by dropping a perpendicular line through the lines of equal hydraulic head (see the arrow on Figure 1.16).

(8) Additional flow directions throughout the map region can be determined by joining two of the three wells in a triangle to an adjacent well with straight lines and repeating the procedure of constructing triangles until all wells of the site have been connected into a net of polygons.

(9) Follow Steps 4–7 to estimate the direction of flow in each of the triangular blocks.

7.4 REFERENCE

Heath, R. C., 1987. *Basic Ground-Water Hydrology.* U.S. Geological Survey Water-Supply Paper 2220.

CHAPTER 8

Analytical Procedure for Estimating Rate of Groundwater Flow

8.1 INTRODUCTION TO METHODS

IF the hydraulic properties of an aquifer are known, the average rate of groundwater flow may be estimated from Darcy's law. For a background discussion of Darcy's Law and groundwater flow rate, see Section 1.8.

This chapter describes two methods for solving Darcy equations: the *Electronic Calculator and Worksheet Method* (Section 8.2) and the *Microcomputer and Electronic Spreadsheet Method* (Section 8.3). Figure 8.1 illustrates a water table map of the flyash storage site from the case study in Chapter 2, and Table 8.1 presents pertinent hydrogeologic information about the site. Use this map and information (or your own project information) for application of the methods described in this chapter.

8.2 ELECTRONIC CALCULATOR AND WORKSHEET METHOD

Purpose

To estimate the rate of groundwater flow through an aquifer.

Equipment and Materials

- aquifer properties (e.g., Table 8.1)
- control point data (e.g., Table 6.1)
- water table map (e.g., Figure 8.1)
- electronic calculator

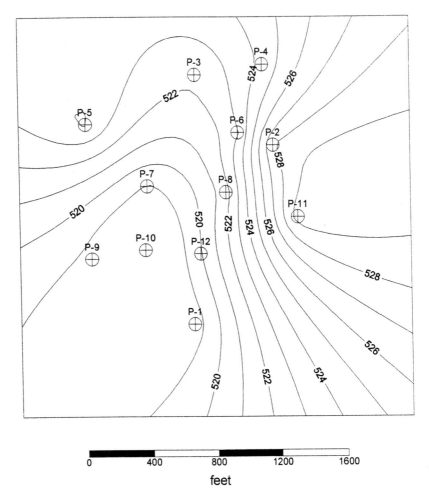

Figure 8.1 Water table base map.

- rate of flow worksheet (e.g., Figure 8.2)
- velocity of flow worksheet (e.g., Figure 8.3)
- pencil (3H)

Procedure

Either English or metric units may be used for the measurements and calculations of this chapter, as long as consistent units are employed throughout.

Defining Groundwater Flow Variables

Determine the elevation of the groundwater level (h_1 and h_2).

(1) Determine the elevation of the water level in two observation wells (e.g., P-11 and P-10 on Figure 8.1) that are sited in a recharge area and a discharge area, respectively (see Table 6.1).

(2) Enter the highest (i.e., max.) water level value in the box labeled h1 and lowest (i.e., min.) value in the box labeled h2 on the rate of flow worksheet (Figure 8.2).

Determine the length of the flow path (l).

(1) Measure the horizontal ground distance between the wells (i.e., the length of the flow path).

(2) Enter this value in the worksheet box labeled l.

Determine the aquifer properties.

(1) Enter the value of the hydraulic conductivity of the aquifer (see Table 8.1 for example) in the worksheet box labeled K.

(2) Enter the saturated thickness of the aquifer (see, for example, Table 8.1) in the worksheet box labeled b.

(3) Enter the width of the aquifer section in the worksheet box labeled w. (If no other information is available, use 1 foot or meter.)

Calculating Flow Values

Follow the key functions illustrated on the rate of flow worksheet (Figure 8.1) to perform the calculations described below and record all interim solutions in the appropriate display boxes.

Calculate the hydraulic head.

(1) Enter the value of the max. water level (h_1).

(2) Press [−].

(3) Enter the value of the min. water level (h_2).

TABLE 8.1. Hydrogeologic Information for the Flyash Storage Site of the Case Study.

Rock type	Fractured, silty black shale
Porosity	20 percent (0.20)
Aquifer type	Unconfined
Hydraulic conductivity	14 ft./day
Saturated thickness	300 feet

Parameter	Operation			Units
max. water level	ENTER	h1		ft or m
	PRESS		-	
min. water level	ENTER	h2		ft or m
	PRESS		=	
HYDRAULIC HEAD	Display	h1 - h2		ft or m
	PRESS		÷	
length of flow path	ENTER	l		ft or m
	PRESS		=	
HYDRAULIC GRADIENT	Display	(h1- h2)/l		dimensionless
	PRESS		X	
hydraulic conductivity	ENTER	K		ft or m /day
	PRESS		=	
DARCY VELOCITY	ENTER	Vdf		ft or m /day
	PRESS		X	
saturated thickness	ENTER	b		ft or m
	PRESS		=	
	PRESS		X	
width of aquifer section	ENTER	w		ft or m
	PRESS		=	
RATE OF FLOW	Display	Q		ft^3 or m^3 /day

Figure 8.2 Rate of flow worksheet.

(4) Press [=].

(5) Enter the result in the display box labeled h1-h2.

Calculate the hydraulic gradient.

(1) Press [÷].

(2) Enter the value of the length of flow path (*l*).

(3) Press [=].

(4) Enter the result in the display box labeled (h1-h2)/l.

Calculate the Darcy velocity.

(1) Press [×].

(2) Enter the value of the hydraulic conductivity (*K*).

(3) Press [=].

(4) Enter the result in the display box labeled Vdf.

Estimating the Rate of Groundwater Flow

Calculate the flow rate.

(1) Press [×].

(2) Enter the value of the saturated thickness of the aquifer (*b*).

(3) Press [=].

(4) Press [×].

(5) Enter the value of the width of the aquifer section (*w*).

(6) Press [=].

(7) Enter the result in the display box labeled Q on the rate of flow worksheet.

Estimating the Flow Velocity

Transfer the groundwater flow variables to the velocity of flow worksheet.

(1) Copy the values of *Vdf* and *l* from the rate of flow worksheet (Figure 8.2) to the appropriate boxes on the velocity of flow worksheet (Figure 8.3).

(2) Enter the porosity of the material, in decimal form (see Table 8.1 for example), in the worksheet box labeled n.

Calculate the flow velocity.

(1) Enter the value of the Darcy velocity (*Vdf*).

(2) Press [÷].

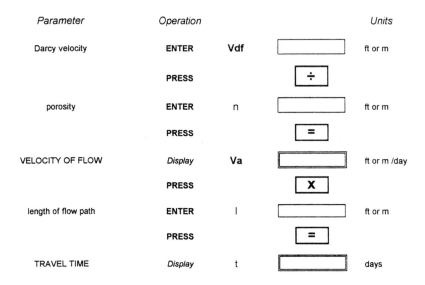

Figure 8.3 Velocity of flow worksheet.

(3) Enter the value of porosity (n).

(4) Press [=].

(5) Enter the result in the display box labeled Va.

Calculate travel time.

Assuming that an upgradient well (e.g., Figure 8.1, well P-11) represents a point source of contamination, calculate the time (i.e., travel time, t) it will take for the contaminant to reach a downgradient well (e.g., Figure 8.1, well P-10), where travel time is the product of the length of flow and the flow velocity ($t = l \times Va$).

(1) Press [×].

(2) Enter the value of the length of flow path (l).

(3) Press [=].

(4) Enter the result in the display box labeled t.

8.3 MICROCOMPUTER AND ELECTRONIC SPREADSHEET METHOD

Purpose

To estimate the rate of groundwater flow through an aquifer.

Equipment and materials

- aquifer properties (Table 8.1)
- control point data (Table 6.1)
- water table map (Figure 8.1)
- microcomputer with an 80386 processor or higher, MS-DOS 3.3 or higher, Windows 3.1 or higher, hard drive, floppy disk drive, and graphics printer
- Windows version of electronic spreadsheet software (e.g., Lotus 1-2-3, Quattro Pro, or Excel)
- Hydrodata Diskette

Procedure

This method employs electronic spreadsheet software to calculate the groundwater flow rate. Map measurements (use Table 8.1 or your own project data) are entered into a spreadsheet template (filename: flowtemp.txt), which is loaded from the Hydrodata Diskette. The calculations of flow parameters are accomplished by means of formulas entered and stored in the cells of the template. When the calculations are completed, a table of the calculations may be printed.

Starting Up

Turn on the computer, monitor and printer. Wait until the Windows desktop is displayed.

Open the application.

(1) Open the Windows group icon that contains your software application (e.g., Quattro Pro), or click on the Windows 95 Start button and point first to Programs and then to the application folder.

(2) Double click on the application icon (e.g., Quattro Pro) from the group window, or click on the name of the application from the drop-down list.

Opening the Flow Data Template File

(1) Insert the Hydrodata Diskette into drive A (or B).

(2) Choose File | Open. The Open File dialog box will be displayed on the screen.

(3) Click on the arrow next to the file type list box and select text files (e.g., *.txt) from the drop-down list.

(4) Specify the drive that contains your file by clicking on the arrow next to the Drives list box and then clicking on the drive name (e.g., A).

(5) Specify the file you want to open by clicking on the name of the desired file in the file name drop-down box (e.g., flowtemp.txt).

(6) Initiate file opening.

For Quattro Pro and Lotus 1-2-3 click on OK.

For Excel: The Text Import Wizard will appear. Click twice on Next > and once on Finish.

The Rate and Velocity of Groundwater Flow Worksheet will be displayed in the spreadsheet window.

Entering Spreadsheet Formulas

For Microsoft Excel users, be sure to enter an equal sign (=) in front of all formulas given below.

(1) Select cell E12.

(2) Type (E4-E5)/E6.

(3) Select cell E13.

(4) Type (E12*E7).

(5) Select cell E14.

(6) Type (E8*E9)*E13.

(7) Select cell E15.

(8) Type (E13/E10).

(9) Select cell E16.

(10) Type (E15*E6).

Defining Groundwater Flow Variables

For the following steps either English or metric units may be used for the measurements and calculations, as long as consistent units are employed throughout.

Determine the elevation of the groundwater level (h_1 and h_2).

(1) Determine the elevation of the water level in two observation wells (e.g., P-11 and P-10 on Figure 8.1) that are sited in a recharge area and a discharge area, respectively (see Table 6.1).

(2) Enter the highest (i.e., max.) water level value in cell E4.

(3) Enter the lowest (i.e., min.) value in cell E5.

Determine the length of the flow path (l).

(1) Measure the horizontal ground distance between the wells (i.e., the length of the flow path).

(2) Enter this value in cell E6.

Determine the aquifer properties. If you are using the properties from the case study in Chapter 2, then have Table 8.1 ready for reference.

(1) Enter the value of the hydraulic conductivity (K) of the aquifer (see Table 8.1, for example) in cell E7.

(2) Enter the value of the saturated thickness of the aquifer (b) in cell E8.

(3) Enter the value of width of the aquifer section (w) in cell E9.

(4) Enter the value of the porosity (n) in cell E10.

Calculating Flow Values

The values of the groundwater flow parameters (hydraulic gradient, Darcy velocity, rate of flow, flow velocity, and travel time) will be calculated and displayed automatically in the box at the bottom of the Rate and Velocity of Groundwater Flow Worksheet.

Saving the Rate and Velocity Data File

The rate and velocity data should be saved on disk for subsequent retrieval and printing. *Warning!* Do not save the data file under the name of the template file or the original template will be overwritten.

Save the new file for the first time.

(1) Select cell A1.

(2) Choose File | Save As. The Save File dialog box will appear on the screen. The name of an original file (e.g., flowtemp.txt) will be displayed and highlighted in the File Name text box.

(3) Press BACKSPACE to erase the name of the template file.

(4) Type the new file name (e.g., a:\flowrate) in the File Name text box.

(5) Click on the arrow next to the File Type text box and select the type of file used by your particular application. (Do not use *.txt.)

(6) Click on OK. The new file will be saved to the drive and directory you specified. Later, to save the existing file choose File | Save. The changes to the file will be saved under the original file name and the old data will be overwritten.

Printing the Water Level Data Form

Check the printer configuration

(1) Choose File|Print. The Print dialog box will appear on the screen.

(2) Make sure that your printer is identified as active. (If not, use the printer setup option to select and configure it.)

Edit page settings.

(1) Click on the Page Setup button and, depending on your application, select:

For Quattro Pro: Print Scaling|Print to fit
For 1-2-3 in the Size text box: Fit all to page
For Excel in the Scaling box: Fit to: 1 page

(2) Click on OK.

Preview the print job.

(1) Click on the Print Preview button. A full-page view of the worksheet will be displayed. To view details, use the zoom option.

(2) Press ESC until the screen returns to the worksheet.

Print the data worksheet.

(1) Choose File|Print. The Print dialog box will be displayed. The selected block appears in the Print or Print What box.

(2) Click on Print (or OK). The Rate and Velocity of Groundwater Flow form should now print. If the format of the printout requires modifying, or if an error in data or labels is evident, return to the worksheet and make the necessary corrections. (The application user's guide or reference manual may be useful for explaining procedures.)

Quitting the Spreadsheet Program

Before closing an application, be sure to save all active files.

(1) Choose File|Exit. If you have saved all active files, the application window closes. If you changed an active file but did not save it, then the Exit (or Save) dialog box will appear.

(2) If the dialog box appears, choose Yes (and, if necessary, Replace) to save the file. The Windows desktop will appear on the screen.

(3) Open another application or quit Windows and turn off computer. Be sure to close all open applications before quitting Windows.

PART 2

GROUNDWATER INVESTIGATIONS

CHAPTER 9

Principles of Groundwater Investigations

9.1 AN OVERVIEW OF GROUNDWATER INVESTIGATIONS

GROUNDWATER investigations are conducted for a variety of purposes. One chief purpose is the exploration for groundwater supplies. Groundwater is the source of nearly one half of the water used in the United States exclusive of hydroelectric generation, and therefore constitutes an important natural resource. A second important purpose is assessment of groundwater contamination. Recent environmental legislation has stimulated increased interest in the fate and transport of contaminants in groundwater.

Groundwater investigations may be classified roughly into two groups: *regional investigations* and *site investigations.* The major difference is the scale on which the investigation is conducted, although some differences in time, cost, and techniques also exist.

A regional groundwater investigation may cover hundreds or even thousands of square miles. It is usually a reconnaissance study which is employed to achieve an over-all evaluation of the hydrogeologic situation. It may be conducted in order to gain an understanding of the occurrence and availability of groundwater within a region, or it may be carried out for the purpose of identifying and describing sources (existing or potential) of contamination. Typically, objectives of regional investigations include the delineation of the important aquifers present, general types and properties of the earth materials that comprise the aquifers, generalized directions of groundwater flow, major sources and rates of recharge and discharge, and chemical quality of the water. Because of time and cost limitations, regional investigations usually rely on existing information available from published sources or records of public agencies.

A site investigation may encompass an area ranging from less than an acre to tens of square miles. Although many of the objectives are the same as those of a regional investigation, a site investigation is more detailed and commonly more complex, and it typically includes intensive field examination, sampling, and testing.

The ultimate purpose of many groundwater investigations is to serve the practical aims of the client—the person, corporation, or agency—who ordered the project. In such a context the investigator, who may be a single person or a team, is responsible for solving hydrologic problems that may originate simply as a request for information, such as "How can I obtain a groundwater supply of 500,000 gallons per day?" or "What is the extent of contamination of the aquifer?" The investigator defines the aims of the project and what the finished product will be. Toward this end, the investigator typically carries out the tasks given below:

- Identify the problems that have to be solved in order to achieve the aims of the project.
- Define the data requirements for solving the problems.
- Evaluate available data and information.
- Select the methods and techniques to be employed in the project.
- Design a work program and schedule.
- Conduct the field and laboratory studies.
- Analyze and interpret the results of the studies.
- Prepare the final report and transmit it to the client.

The problems to be solved depend on the purpose of the investigation. For example, the ultimate goal of a groundwater exploration program may be to locate one or more wells from which a specified amount of water can be produced. In order to accomplish this goal, several specific hydrogeologic problems must be solved. These include

- identification of potential aquifers
- nature and origin of the permeability of aquifers
- geometric configurations of aquifers
- boundaries of groundwater systems
- position of the water table (or potentiometric surface)
- depth of drilling required
- estimate of supply and replenishment
- water quality
- constraints on the location of wells

In order to solve the problems listed above, a hydrogeologist may require most or all of the following information:

- location of the boundaries of the aquifers
- lithologic composition of aquifers

- thickness of the aquifers
- expected well yield of aquifers
- location and configuration of land forms and surface water bodies
- location and geometry of fracture traces
- depth to and configuration of water table
- water budget
- chemical composition of groundwater
- potential sources of contamination
- location of sites for exploratory test wells

The preliminary work of an investigation usually begins with a compilation and review of all available hydrogeologic information on the study region. This information is obtainable from

- topographic maps
- geologic reports and maps
- hydrologic reports and maps
- soil surveys
- aerial photographs and other remote images
- well drillers' records

Some common sources of maps, reports, aerial photos, and related background information include

- U.S. Geological Survey
- U.S. Soil Conservation Service
- U.S. Army Corps of Engineers
- U.S. National Climatic Center
- U.S. National Oceanographic and Atmospheric Administration
- U.S. Environmental Protection Agency
- EROS Data Center
- state geologic and hydrologic surveys
- colleges and universities

The precise methods of investigation vary according to the aims and nature of the particular project. Preliminary conclusions concerning the regional occurrence of groundwater can be obtained by means of geologic maps, aerial photographs, and ground reconnaissance. These exploratory tools reveal important information about the types of rock present in the region, the position and thickness of water-bearing strata, the presence (or absence) of fracture zones, and topographic or surficial features. Additionally, site investigations require detailed geologic mapping and rock, soil, and water sampling. These field techniques are commonly suplemented with geophysical surveys.

Depending on the scale and particular purpose of the investigation, the work progam is divided into two phases: 1) characterization of the hydrogeologic setting and 2) location of test or monitoring wells. The results of the first phase provide the information that is necessary for the effective execution of the second phase.

Field and laboratory studies are conducted in order to obtain any information about the project region that is not available from reports, maps, etc. and to obtain detailed data about specific sites within the region. At the very least, a field reconnaissance of the region should be carried out for the purpose of verifying preliminary information, such as the presence of particular rock types or fracture traces.

The data compiled during the investigation must be analyzed and interpreted in accord with the aims of the project. For a groundwater supply investigation, a client wants to know where to drill and how much water will be obtained. For a groundwater contamination investigation, a client wants to know what kind of contaminants are present, how much is present, and where it is found. The data that support the answers to these questions should be displayed on tables (e.g., well yields), well or boring logs, maps (e.g., geologic, groundwater availability, and test well maps), and graphs of various kinds.

Although the precise organization and contents of a final project report may vary depending on the nature of the project, most will follow the general outline below (from Fetter, 1994, p. 589).

(1) A title page describing the report, who prepared it, for whom it was prepared, and a date
(2) An introduction stating the purpose of the study, why it was carried out, and the general conditions under which it was conducted
(3) Conclusions that can be drawn about the region or site on the basis of the study
(4) Recommendations to the client regarding the use of the site or the need for additional studies
(5) The body consisting of
   • a review of previous work
   • a description of the procedures and methods used in the study
   • the general results of the field study and laboratory analyses
   • an interpretation of the findings
   • appropriate exhibits (e.g., tables, figures, plates, etc.)
(6) An appendix consisting of
   • acknowledgements of assistance given by others during the study
   • a bibliography
   • technical data, computations, and other supporting evidence

9.2 CHARACTERIZING THE HYDROGEOLOGIC SETTING

Because geologic materials strongly influence the location and availability of groundwater and the nature and extent of groundwater contamination, full knowledge of the type of materials, their areal distribution, their stratigraphic relationships, and their structural characteristics is paramount in groundwater exploration and contamination assessment. As Freeze and Cherry (1979, pp. 10–11) state, "The groundwater hydrologist or geologist must have some background in the interpretation of geologic evidence and some flair for the visualization of geologic environments."

The term "hydrogeologic setting" refers to a composite description of the major geologic and hydrologic factors that influence the occurrence and movement of groundwater. The hydrogeologic setting of a region or a site may be interpreted and illustrated by effectively using the following techniques and exhibits:

- topographic maps
- geologic maps, columns, and structure sections
- soil, rock, and water samples
- fracture trace maps, prepared from aerial photographs or radar images
- hydrogeologic maps
- geophysical surveys
- water well records
- aquifer tests

Because of the relationship that exists between groundwater and topography (see Section 1.5), a topographic map is a useful tool for any groundwater investigation. In addition to illustrating the cultural and natural features of a region, topographic maps provide important information on surface-drainage patterns and gradients, recharge and discharge areas, and even estimates of depth to groundwater and direction of groundwater flow. Further, they may serve as useful base maps on which geologic and hydrologic information can be displayed.

A geologic map (see Figure 9.1) displays the distribution of rock formations at the surface of the earth by various colors or patterns. It provides important information on the areal extent, thickness, composition, and structure of the earth materials of a region or site. Surficial geology maps illustrate the kind and distribution of the unconsolidated material (e.g., sand and gravel deposits) overlying bedrock. Bedrock geology maps display the kind and distribution of bedrock formations and may also show the location of geologic structures, such as fold axes and fault traces.

A geologic column (or columnar section) illustrates the stratigraphic sequence and thickness of the formations of a region. Following the Principle of Superposition, the youngest rock unit is shown at the top of the column and the oldest at the base (see Figure 9.2).

A structure section (or geologic cross section) is a two-dimensional dia-

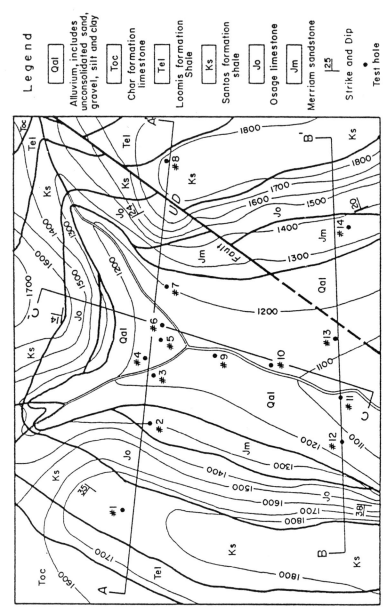

Figure 9.1 Example of a geologic map (from Gibson and Singer, 1971, p. 29).

104

Sys-tem	Series	Formation	Sym-bol	Columnar Section	Thickness in Feet	Character and Distribution
Quaternary		Bolson deposits	Qb		700+	Gravel, sand and clay in Salt Flat. Unconsolidated impure gypsum at the surface in lowest part of basin.
		—UNCONFORMITY—				
Cretaceous	Lower Cretaceous	Comanche series	Kc		400±	Buff sandstone, with subordinate amount of conglomerate, shale, and limestone. Caps the hills 6 miles west of Van Horn and in the vicinity of Plateau station.
		— UNCONFORMITY —				
		Rustler limestone	Cr		200±	Fine-grained gray to whitish magnesian limestone in faulted area in eastern part of quadrangle.
Carboniferous	Permian	Castile gypsum	Cc		275±	Massive-bedded gypsum in faulted area in eastern part of quadrangle.
		—UNCONFORMITY—				
		Delaware Mountain formation	Cd		2000 +	Interbedded gray limestone and buff sandstone in Delaware Mountains; massive white and gray limestone member in Apache Mountains.
	Pennsylvanian	SEQUENCE CONCEALED Hueco limestone	Ch	SEQUENCE CONCEALED	2500+	Massive gray limestone with basal conglomerate. In Sierra, Diablo, Baylor, Beach, Wylie, and Carrizo Mountains.

Figure 9.2 Example of a geologic column (reprinted from Lahee, 1961, p. 725, with permission).

gram that represents a vertical slice through the upper part of the crust of the earth (see Figure 9.3).

Soil, rock, and water samples disclose detailed information about the hydrogeologic materials of a region or site which cannot be obtained by indirect means such as maps. Although a regional groundwater investigation may not include sampling, no site investigation should be conducted without reliance on samples. Samples of bedrock may be collected directly from outcrops present at a site. Shallow soil samples may often be collected by using a hand soil auger; however, relatively deep soil and bedrock samples require the use of power augers or drills. The geologic

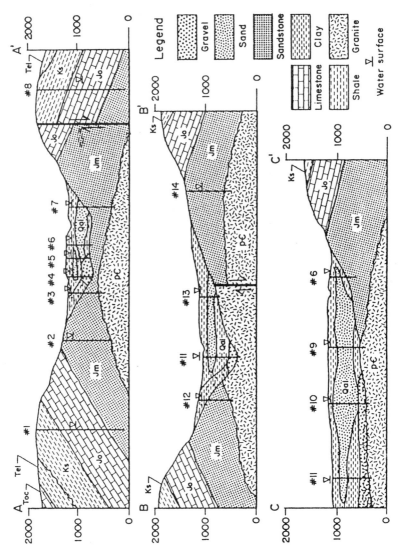

Figure 9.3 Structure sections from geologic map of Figure 9.1 (from Gibson and Singer, 1971, p. 30).

106

character of the materials (and other kinds of hydrogeologic information) encountered at measured depths and intervals are displayed on a lithologic log (see Figure 9.4). The collection of water samples, whether from springs and seeps or from wells, provide important information on the natural background quality of the groundwater and on any contaminants present.

Aerial photographs, and often radar images, are essential elements of any groundwater investigation. They may reveal hydrogeologic information that cannot be discerned on maps or observed clearly on the ground. In addition to determining the location of cultural features, they are useful for identifying springs, refining the boundaries of geologic units, and determining the trends and sizes of fracture traces and lineaments.

A fracture trace map (see Figure 9.5) illustrates the locations of fracture traces identified on aerial photographs of a region. Fracture traces "visible on aerial photographs are natural linear-drainage, soil tonal, and topographic alignments which are probably the surface manifestations of underlying zones of fracture concentration" (Lattman and Parizek, 1964). These fracture zones are less resistant to erosion than the rock block which lies between them (see Figure 9.6); hence, in the field, fracture traces may be visible as elongated surface depressions or straight stream segments. If they are zones of concentrated groundwater discharge, there may be a line of springs along them. In carbonate terrains, aligned sinkholes are a typical surface expression. Such zones of fractures are capable of transmitting large quantities of groundwater and therefore provide promising sites for water supply wells. Natural linear features that range from 300–1,500 meters long are considered to be fracture traces. Those longer than 1,500 meters are termed lineaments and may reach lengths of 150 kilometers. [For additional information, see Lattman (1958), Lattman and Parizek (1964), and Meiser & Earl, Hydrogeologists (1982).]

A hydrogeologic map resembles a geologic map in that rock units are illustrated by colors or patterns, but various kinds of hydrologic information are also displayed. For example, a hydrogeologic map may show the location of water wells, reported well yield, the potentiometric surface for each aquifer, and chemical characteristics of the groundwater. One common type of hydrogeologic map is a groundwater availability map, which exhibits information about the quantity of water potentially available from the aquifers of a region (i.e., expected yield).

Most groundwater exploration projects can profit from geophysical surveys. Surface geophysical surveys may include such methods as seismic reflection and refraction, electrical resistivity, and electromagnetic conductivity. These techniques make use of variations of density, electrical conductivity, magnetic properties, elasticity, and other physical properties of the earth. Some geophysical techniques identify the presence of subsur-

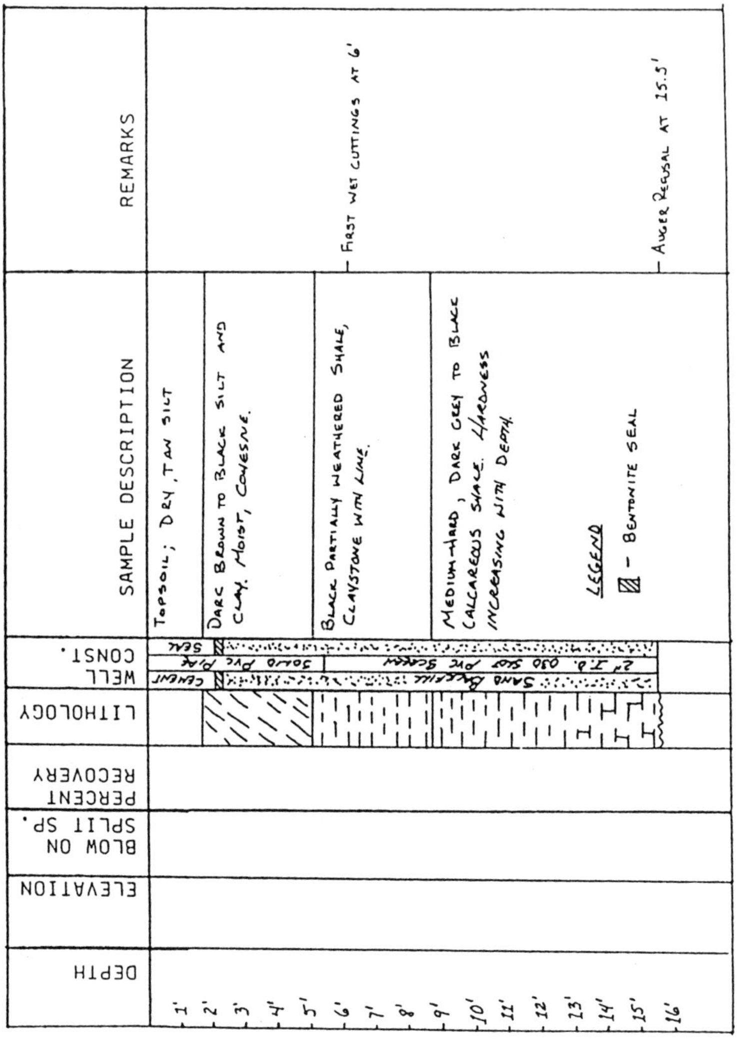

DEPTH | ELEVATION | BLOW ON SPLIT SP. | PERCENT RECOVERY | LITHOLOGY | WELL CONST.

REMARKS

SAMPLE DESCRIPTION

Topsoil; Dry, Tan Silt

Dark Brown to Black Silt and Clay. Moist, Cohesive.

Black Partially Weathered Shale, Claystone with Lime.

Medium-Hard, Dark Grey to Black Calcareous Shale. Hardness increasing with Depth.

First wet cuttings at 6'

Auger Refusal at 15.5'

Legend

⊠ — Bentonite Seal

Bentonite Seal

Cement

Sand Backfill

2" I.D. .030 Slot PVC Screen

Solid PVC Pipe

1' 2' 3' 4' 5' 6' 7' 8' 9' 10' 11' 12' 13' 14' 15' 16'

Figure 9.4 Example of a lithologic log.

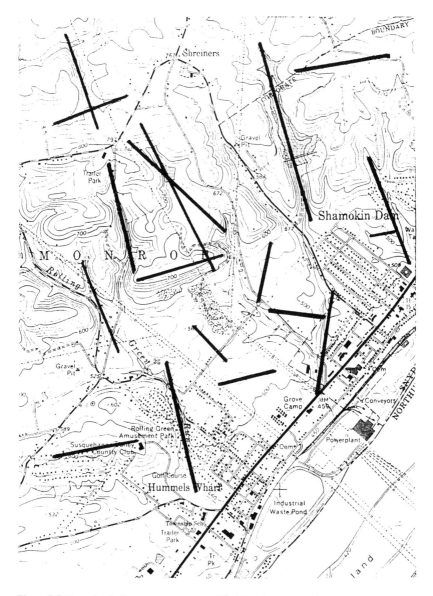

Figure 9.5 Example of a fracture trace map (modified from Lennon and Meyers, 1988, pp. 5–8).

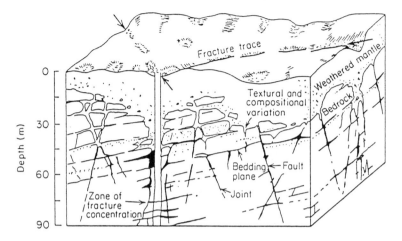

Figure 9.6 Relationship between fracture traces and fracture concentrations. (Reprinted from Lattman and Parizek, 1964, p. 87 with land permission from Elsevier Science – NL, Sara Burgerhartstraat 25, 1055 KV Amsterdam, The Netherlands.)

face water, while others do not; yet all are useful in the interpretation of the shallow subsurface, and they may eliminate or reduce the necessity of an extensive drilling program (see Table 9.1). Two excellent reviews of surface geophysical methods in groundwater investigations are given by Driscoll (1986, pp. 168–180) and Evans (1982).

Water well records are a valuable source of hydrogeologic information about a region; they include well location, depth, and yield data. The primary source of well records is local well drillers. However, many state geologic and water resources agencies maintain central records of wells,

TABLE 9.1. Complementary Surface Geophysical Methods.

Application	Primary Methods	Secondary Methods
Subsurface geology	Seismic, ground penetrating radar	Electromagnetic conductivity, resistivity
Depth to water	Ground penetrating radar	Electromagnetic conductivity, resistivity
Leachate plumes	Electromagnetic conductivity, resistivity	Ground penetrating radar
Buried metallic wastes	Magnetometry, metal detection	Electromagnetic conductivity, resistivity, ground penetrating radar
Buried nonmetallic wastes	Ground penetrating radar, electromagnetic conductivity	Resistivity

which may be available on computer files or in published summaries. Summaries of well records can provide useful information to the hydrogeologist, including the location of the wells, name of the property owner, name of the driller, topographic setting, rock and aquifer types, well depth, reported well yield, etc. (see Figure 9.7). Reported well yield is the amount of water pumped from a well in a given period of time that lowers and holds the water level near the bottom of the well (Freeze and Cherry, 1979, p. 305). In most cases it is measured by the driller at the time the well is drilled and is commonly reported in units of gallons per minute (gpm).

An estimation of the availability of groundwater in an aquifer, as well as a comparison of water yield from different aquifers of a region, begins with the construction of a frequency distribution of reported well yields. A

A. WATER WELL RECORDS

(Data from drillers' water well completion reports on file at the Pennsylvania Geological Survey and from earlier reports in Lohman, 1938)

Well number: Letter prefix indicates county in which well is located. Number relates record in this table to well's location on Plate 5.
Topographic location: D, drainage divide; H, hilltop; S, slope; V, valley bottom.
Driller: JEH, John Hockenberry, Landisburg, Pa.; JMH, J. M. Hubler, Port Royal, Pa.; K, Kohl Bros., Harrisburg, Pa.; Z, Zechman Bros., Middleburg, Pa.
Accuracy of location: 1, within 100 ft.; 2, within 1,000 ft.; 4, within 1 mile. Locations were furnished by drillers and well owners, and in most cases have not been checked in the field.
Elevation: Estimated from topographic maps.
Aquifer: Mpm, Pocono Formation, Mount Carbon Member; Dcd, Catskill Formation, Duncannon Member; Dcs, Catskill Formation, Sherman Creek Member; Dci, Catskill Formation, Irish Valley Member; Dth, Trimmers Rock and Harrell Formations; Dms, Mahantango Formation, Sherman Ridge Member; Dmm, Mahantango Formation, Montebello Member; Dmf, Mahantango Formation, Fisher Ridge Member; Dmr, Marcellus Formation; Don, Onondaga Formation; SDk, Keyser Formation; Sto, Tonoloway Formation; Sw, Wills Creek Formation; Sb, Bloomsburg Formation; Sm, Mifflintown Formation; Sr, Rose Hill Formation.
Reported lithology of aquifer: Ls, limestone; Sh, shale; Ss, sandstone.
Water use: D, domestic; I, industrial; N, none; P, public water supply; S, stock.

Well number	Township	Well owner	Topographic location	Driller	Accuracy of location	Date completed	Elevation	Casing diameter (inches)	Casing length (feet)	Total depth (feet)	Depth to bedrock (feet)	Depth to water-bearing zones (feet)	Aquifer	Reported lithology of aquifer	Static water level (feet below the land surface)	Reported yield (gpm)	Specific capacity of well (No. of ft. drawdown/yield in gpm)	Water use
							JUNIATA COUNTY											
JU-3	Delaware	J. H. Wirt	V?	JMH	4	3/68	590'	5⅝	34	57	28	40, 51	Sto	Ls	27	14	—	D
JU-13	Delaware	K. Leach	S	Z	1	6/66	650	6⅝	44	147	85	122, 140	Dmm	Sh	—	7	—	D
JU-14	Delaware	Ray T. Benner	V	Z	1	5/67	490	5⅝	102'	190	90	135, 160	Sto	Sh	50	150	—	D
JU-26	Delaware	Zigler	S	Z	1	9/66	595	6⅝	58	172	55	90, 140, 166	Dmm	Sh	35	10	—	D
JU-38	Delaware	Jacob Kauffman	V	Z	1	4/69	550	6⅝	40	122	16	70, 112, 115	SDk	Ls	6	20	—	D
JU-39	Delaware	Gerald Swartz	D	Z	1	4/69	548	6⅝	50	122	73	85, 110	Dmm	Ls	—	7	—	D
JU-40	Delaware	Philip C. Varner	V	Z	1	11/66	555	6⅝	156	197	146	160, 183, 190	Sto	Sh	50	10	—	D
JU-41	Delaware	Harmon Begrade	D	Z	1	12/61	548	6⅝	57	122	50	90, 109, 118	Dmm	Sh	50	6	—	D
JU-101	Delaware	William Gilpin	V	Z	2	1/70	595	6⅝	42	396	35	150, 305, 350	Dms	Sh	—	5	—	D
JU-104	Delaware	J. M. Graybill	V?		4	1934	620	6	22	40			Dmf	Sh, Ss	3	3	—	D
JU-105	Delaware	T. Hubbard	V?		4	1934	650	6	24	74			Dms	Gy Sh	35	1½	—	S
JU-106	Delaware	Ralph Freed	S?		4	1934	700	6	18	55			Dmm	Blk Sh	45	2	—	D
JU-107	Delaware	M. E. Schlegal	V?		4	1934	460	5		45			Sto	Ls	8	3	—	N
JU-108	Delaware	Breyer Ice Cream	V?		4	1934	430	6	20	180		100	Sw	Ls	2	"large"	—	N
JU-109	Delaware	Thompson Water Works	V?		4	1934	640	6		70			Dmm	Blk Sh	0–2	30	1.50	N
JU-4	Fayette	John Scheil	V?	JMH	4	11/67	670	5⅝	20	56	16	31, 44	Sto	Ls	10	15	—	D
JU-6	Fayette	J. H. Wirt	V	JMH	2	3/68	675	5⅝	30	54	21	36, 48	Sto	Ls	20	12	—	D
JU-7	Fayette	Chester Landis	V	JMH	2	9/68	670	5⅝	18	48	12	31, 47	Sto	Ls	20	12	—	D
JU-8	Fayette	Phyllis Hepner	V	JMH	2	1/67	640	6	16	28	10	18, 27	Dmr	Ss	11	20	—	D
JU-36	Fayette	Ken Hepner	S	Z	1	7/68	810	6⅝	60	397	55	211, 363, 393	Sm	Ls	—	4	—	D
JU-37	Fayette	Roy Smith	V	Z	1	11/67	715	6⅝	62	147	55	72, 122, 130	Don	Sh, Ls	50	5	—	D
JU-110	Fayette	J. D. Lefber	V?		4	1934	640	6	20	44			Don	Ls	24	3	—	D
JU-111	Fayette	Mrs. Scholl	V?		4	1934	700		6	94			Sw	Calc. Sh	—	3	—	D
JU-1	Greenwood	Troup Bros. & Benner	V	JMH	1	9/68	575	5⅝	20	74	12	40, 65	Dms	Sh	20	9	1.80	I
JU-2	Greenwood	D. P. Fisher	V	JMH	2	8/68	565	5⅝	20	46	12	32, 44	Dms	iron rock	10	6	—	D
JU-113	Greenwood	Claude Swartz	H?		4	1934	830	6	22	135			Dth	Sdy Sh	100	1–3		D, S
JU-114	Greenwood	Troup Bros.	V?		4	1934	560	6	10	40			Dms	Sh	10	2	—	D
JU-9	Monroe	Adam Fogle	H?	JMH	4	11/67	960	5⅝	33	56	18	45, 50	Dci	Slate	40	7	—	D
JU-10	Monroe	Chas. Bollinger	V?	JMH	4	6/68	700	5⅝	23	60	13	27, 52	Dth	Sh	22	10	1.25	D
JU-11	Monroe	Glen L. Kauffman	V	Z	1	5/68	650	6⅝	43	462	25	206, 413	Dth	Sh	110	2	—	D
JU-15	Monroe	J. A. Strawser	V	Z	1	2/69	635	6⅝	43	122	29	62, 93, 112	Dth	Ls		50	—	D
JU-16	Monroe	J. Beaver	S	Z	2	1/68	840	6⅝	40	247	30	48, 190, 242	Dci	Sh		5	—	D
JU-17	Monroe	G. R. Matthews	V	Z	1	7/67	760	6⅝	40	112	10	61, 103	Dth	Ls	full	80	—	D
JU-18	Monroe	Earl Pyle	H	Z	1	2/67	850	6⅝	42	172	36	47, 95, 158	Dci	Sh	40	5	—	D
JU-21	Monroe	G. Matthews	S	Z	2		760	6⅝	20	97	10	47	Dth	Sh	10	50	—	D

Figure 9.7 Example of well records summary (from Faill and Wells, 1974, pp. 270–271).

frequency distribution is "a list of events being measured for a total population or sample, with a number indicating how many times each event occurs" (Couch, 1982, p. 13). A frequency distribution table displays how many wells of the aquifer yield a given amount of water. In most instances it is useful to construct a lognormal probability graph of well yield distributions (see Figure 9.8). On this type of graph, the relationship between well yield and relative cumulative frequency percent takes the form of a straight (or nearly straight) line. The lognormal plot in Figure 9.8 shows the percentage of wells in which indicated yield was equaled or exceeded. (Some published frequency distribution plots illustrate the percentage of yields equal to or less than the indicated yield.)

Lastly, an important element of the characterization of a hydrogeologic setting is a determination of various hydrologic properties, including hydraulic gradient, direction and rate of flow, hydraulic conductivity, transmissivity, and storativity. (The last two properties are defined and examined in Part 3.) Numerous techniques are available for measuring these properties; a few of the techniques include a) laboratory tests on small, undisturbed samples of the rock or sediment and b) field tests that include pumping tests, slug tests, and tracer tests. Pumping tests are commonly employed to determine the transmissivity and storativity of aquifers. Slug tests are conducted to determine hydraulic conductivity. Tracer tests can be used to discover if two locations are hydraulically connected and to measure flow velocities. For more information on aquifer tests, see Part 3.

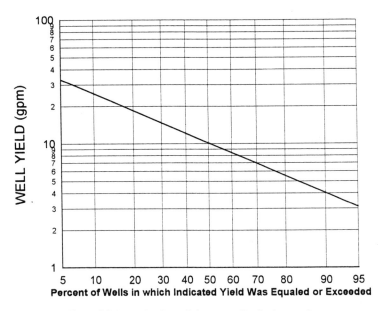

Figure 9.8 Example of a well frequency distribution graph.

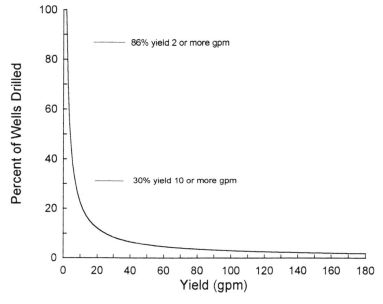

Figure 9.9 Yield of bedrock wells in coastal Maine (modified from Caswell, 1979, p. 80).

9.3 SITING WATER SUPPLY WELLS

According to Meiser and Earl (1982, p. 1), "There are many reasons to drill a well, including water supply, ground-water monitoring, cooling and heating water, and waste injection." All of these reasons carry the implicit need for a well to have a yield high enough for the intended use. In the case of water supply wells, "high enough" depends largely on whether the well is to be used for household (that is, domestic) requirements or for commercial or industrial purposes.

As Davis and DeWiest (1966, p. 260) point out, "Most water wells are located, drilled, and tested without the aid of professional hydrogeologists." They are sited randomly or for the convenience of the property owner or well driller. The chief reason for this is that the cost of scientifically based studies commonly exceeds the expected benefits, a factor pertinent to shallow or low yield (e.g., domestic) wells in particular. For example, Caswell (1979, p. 78) shows that nearly nine out of ten bedrock wells in Maine yield two gallons per minute or more (see Figure 9.9), which is usually high enough for their intended use as a household supply.

For wells that must produce commercial or industrial quantities of water, however, the professional guidance of a hydrogeologist is warranted. An unsuccessful well—one with too low a yield or, indeed, even a "dry hole"—may constitute a serious financial loss to the property owner.

Davis and DeWiest (1966, p. 328) note that, "Monetary savings can be achieved with professional help by the proper location of wells for maximum yield, the specification of total depths of wells, and the calculation of the optimum pumping rate possible for wells." As Figure 9.9 demonstrates, a few wells of any particular region are able to produce more than 50 gpm. This fact indicates that highly permeable zones exist in most bedrock regions and that the appropriate hydrogeologic methods might be used to improve the chances of locating water in these zones.

Well yield in any region may be highly variable and is the consequence of numerous factors, having to do with both the hydrogeologic environment of the well and the specifications of the well itself. For example, Caswell (1979, p. 62) reports that whereas the average well yield from most bedrock settings in Maine is less than 10 gpm, the yield for 6-inch diameter domestic wells ranges from 0 gpm to at least 300 gpm. Commercial and industrial wells, of typically larger diameters, may yield more than 1,000 gpm, however.

The chief factors that influence well yield in bedrock (e.g., crystalline rocks of consolidated sedimentary rocks) are

- fracture spacing
- rock type
- topographic location
- structural setting
- well depth
- well diameter

It is clear that many of these are not independent variables. For example, rock type and fracture spacing are related to one another. Likewise, in a given geomorphic and climatic setting, rock type and topographic expression are related. Regrettably, current information about these factors does not yet permit the strict and direct correlation between any one factor and well yield.

The permeability of a bedrock aquifer—and thus well yield—is determined by the presence (or absence) of fractures, such as faults and joints. Caswell (1979, p. 81) states, "It is common in crystalline rocks to find a high-yield well only a few hundred feet from a low-yield well, or even a dry hole. The productive well intersects one or more water-bearing fractures, while the nonproductive well intersects no fractures, or only low-yield fractures. It is possible in places where the water-bearing fractures are close to vertical that a well can be drilled and not intersect any fractures" (see Figure 9.10). In many regions of crystalline rock most joints are nearly vertical, and the spacing between joints will be generally 0.5–10 feet, that is, equal to or greater than the diameter of most drilled wells. As a consequence, a vertical well may intersect only one or two joint planes.

(a) WIDELY SPACED, NEARLY VERTICAL FRACTURES, NONE OF WHICH ARE INTERSECTED BY THE DRILL HOLE. HOLE IS DRY.

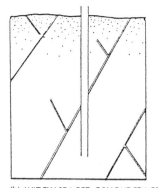

(b) WIDELY SPACED OBLIQUE FRACTURES OCCASIONALLY INTERSECTED BY A SECOND SET OF OBLIQUE FRACTURES. SMALL YIELD.

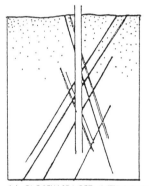

(c) CLOSELY SPACED, INTERCONNECTED FRACTURES. HIGH YIELD.

Figure 9.10 Relationship between well yield and fracture spacing (from Caswell, 1979, p. 87.)

Because of the weight of overlying rock, fractures generally decrease with depth below the land surface. As Caswell (1979, p. 64) observes, "Most water-bearing fractures are likely to die out at depths below about 500 feet, according to data obtained in the coastal region of Maine." Figure 9.11 illustrates that a sharp decrease in the frequency of water-bearing fractures appears to occur at approximately 500 feet in depth. Caswell concludes that drilling "beyond 500 feet in search of water is unlikely to be successful."

The quantities of groundwater available—and thus well yield—are highly variable from one rock type to another. For example, gravel and sand are permeable and serve as good sources of groundwater, whereas

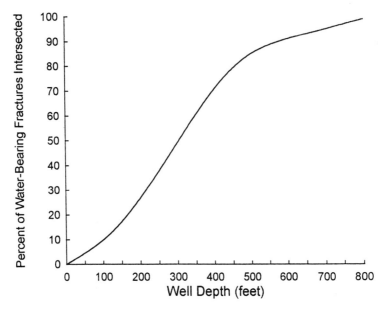

Figure 9.11 Change in percentage of water-bearing fractures intersected with depth (modified from Caswell, 1979, p. 65).

clays exhibit low permeability and, thus, usually make poor sources of water. Likewise, the quantities of water available from crystalline and consolidated sedimentary bedrock varies greatly, depending chiefly on the fracture permeability. Johnston (1970, p. 64) observes that limestones containing fractures enlarged by solution typically yield more groundwater than most other kinds of bedrock. Data from Faill and Wells (1974, p. 187) support this observation (see Table 9.2).

Well yield varies also with topographic location of the well. Several studies of regions that are underlain by bedrock have shown that the average yield of wells in valleys is greater than that of wells situated at topographically higher sites. Table 9.3 shows a comparison between average

TABLE 9.2. Comparison of Average Water Well
Yield with the Reported Rock Type (after Faill
and Wells, 1974, p. 187).

Rock Type	Average Yield (gpm)	Number of Wells
Limestone	16.6	26 (23%)
Shale	15.8	74 (65%)
Sandstone	15.2	13 (11%)

TABLE 9.3. Comparison of Average Yield
with Topographic Location (after Faill
and Wells, 1974, p. 187).

Well Location	Average Yield (gpm)	Number of Wells
Hilltop	8	3 (4%)
Drainage divide	20	4 (5%)
Slope	14	23 (30%)
Valley	28	47 (51%)

well yield and four categories of topographic location in the Valley and Ridge of central Pennsylvania. These data illustrate that the average yield of valley wells is 3.5 times greater than that of hilltop wells. Campbell and Lehr (1973, p. 247) report a similar relationship between well yield and topographic location in the crystalline rocks of North Carolina (see Table 9.4). In their study of carbonate rocks in the Valley and Ridge of central Pennsylvania, Siddiqui and Parizek (1971) found that the average yield of valley-bottom wells was eight times higher than that of upland wells. The yield of valley-wall wells fell between these two extremes.

Johnston (1970, p. 66) explains that the average permeability of the bedrock beneath valleys is higher than that beneath hillsides and hilltops. "Where permeability differences related to topographic position occur in areas underlain by the same or similar rock types, the differences may be related to differences in the volume of water that passes through the rocks. As ground water moves from drainage divides toward streams, the volume of water increases because of the addition of recharge from precipitation; thus, a progressively larger volume of water must pass through the same — or even smaller — volume of rock as the ground water approaches the stream." Thus, he concludes, "Valleys are favorable areas for drilling

TABLE 9.4. Relationship between Well Yield and Topographic
Location (after Campbell and Lehr, 1973, p. 247).

Topographic Location	Number of Wells	Average Depth (ft.)	Average Yield (gpm)	Yield per Foot of Well (gpm)	Percent of Wells Yielding Less than 1 gpm
Hill	282	147	7.5	0.052	28
Flat	152	154	17	0.113	3
Slope	228	127	14	0.108	6
Draw	66	180	27	0.148	3
Valley	74	212	28	0.132	1
All wells	802	151	14.5	0.097	12.5

wells because of the high average bedrock permeabilities, and because of the potential for obtaining induced recharge from streams" (Johnston, 1970, p. 66). Davis and DeWiest (1966, p. 328) note also that the highest well yields may occur in or close to broad ravines. "Ravines are developed in many places along permeable fault zones, which explains the higher yield of wells."

The structural setting of a well site also influences well yield. For example, Siddiqui and Parizek (1971) found that both the proximity of water wells to fold axes and the angle of dip of strata influence well yield. They concluded that, in general, wells situated near anticlinal axes (and perhaps synclinal axes) produced higher yields (see Figure 9.12, wells A and C). This phenomenon is related to the presence of fracture concentrations (and, in the case of carbonates, to solution enlargement) in the region of fold axes. Furthermore, wells in consolidated sedimentary rocks having shallow bedding dip angles (i.e., less than 15 degrees) have higher yields than those in more steeply dipping rocks. In carbonate rocks, increased solution occurs in strata with shallow dip angles; yet another variety of rock layers are intersected by wells in gently dipping strata than in steeply dipping strata.

Numerous studies have shown a relationship between well yield and well depth. In bedrock wells, well depth exhibits a direct correlation with the reported depth to shallowest water producing zones. Faill and Wells (1974, p. 186), in a study conducted in the Valley and Ridge of central Pennsylvania, report that most wells "are less than 150 feet (46 m) deep and that water-bearing zones are generally encountered at less than 100 feet (30 m)." Johnston (1970) found that most of the yield of wells may be expected to come from depths of less than 200 feet. Caswell (1979, p. 62)

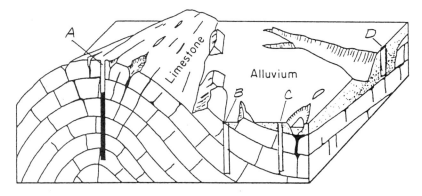

Figure 9.12 Relationship between structural setting and well yield (reprinted from Davis and DeWiest, 1966, p. 363, with permission).

TABLE 9.5. Relationship between Well Diameter
and Well Yield (adapted from
Driscoll, 1986, p. 445).

Well Diameter (in.)	Well Yield (gpm)
6	100
12	110
18	117
24	122
36	131
48	137

states, "Bedrock wells obtain ground water from water-bearing fractures; thus, a well is drilled to a finished depth at which at least one water-bearing fracture is encountered. Records for wells in the mid-Coastal part of Maine suggest that the first such fracture is struck most often before drilling 100 feet, and that there is a 90% chance of encountering water before drilling 200 feet." Hence, for domestic wells at least, enough water to supply the needs of a family home is usually encountered at relatively shallow depths; drilling deeper is seldom productive.

It may seem intuitively obvious that well yield is influenced by the diameter of the well, with larger wells yielding more water than smaller ones; however, for both theoretical and empirical reasons, increasing well diameter does not significantly increase well yield. Table 9.5 shows an example of a 6-inch well that yields 100 gpm with a certain amount of drawdown. If a 12-inch well is constructed at the same site, this well will yield 110 gpm at the same drawdown. A 48-inch well will yield 137 gpm, or 37% more water, at the same drawdown. It is clear, then, that doubling the diameter of the well may be expected to increase yield only by approximately 10% when other factors are unchanged. [For additional information, see Davis and DeWiest (1966), pp. 318–417.]

As the discussion of the factors that influence well yield in the previous section shows, the proper siting of a well can optimize the yield. Figure 9.13 illustrates a hypothetical region that shows the relationship between topographic-geologic features and expected well yield. The highest well yields are produced from those wells which intersect intensively fractured zones (Wells 1 and 4), cavernous rock (Well 6), or are drilled in alluvium (Wells 4 and 8). Well 2 is an example of a low-yield well which is drilled on a hillslope and does not intersect fractured bedrock. Well 1 is a hilltop well which would have a low yield except that it intersects a fault zone.

Drilling sites for water supply wells are selected on the basis of both planning (or engineering) considerations and hydrogeologic consider-

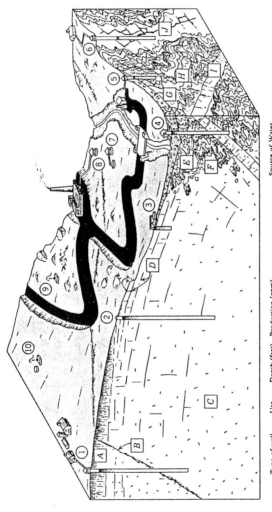

Type of well	Use	Depth (feet)	Production (gpm)	Source of Water
1. Drilled	Farm	210	25.0	Lower part of weathered granite and fault zone. Small amount from joints.
2. Drilled	None	200	0.1	Very small amount from joints.
3. Drilled	Stock	30	0.5	Small amount from joints. Water is artesian.
4. Drilled	Observation	125	15.0	Lower part of alluvium and fractures and joints in and near dike.
5. Drilled	Domestic	80	1.5	Lower part of colluvium and joints in schist.
6. Drilled	Domestic	130	45.0	Cavernous zone in small body of marble.
7. Dug	Stock	20	4.5	Alluvium.
8. Drilled	Industry	160	35.0	Lower part of alluvium and same fault as in well #1.
9. Dug	None	15	0.2	Small amount from joints. Well dry during droughts.
10. Dug	Stock	25	0.7	Weathered granite.

GEOLOGIC UNITS

A. Residual soil on granite.　C. Granite.　E. Alluvium.　G. Colluvium.　I. Aplite dike.

B. Fault.　D. Joints in granite.　F. Contact between granite and schist.　H. Schist.　J. Marble.

Figure 9.13 Relationship between topographic-geologic features and expected well yield (modified from Davis and DeWiest, 1966, p. 329).

ations. Hydrogeologically, the sites that tend to produce the highest well yields may be identified by careful evaluation of factors listed below:

- fracture trace
- expected well yield
- topographic setting
- structural setting
- water quality
- thickness of aquifer
- depth to water
- extent of recharge area
- potential sources of contamination

Test wells should be drilled at sites that maximize the hydrogeologic advantages of these features. Despite a thorough knowledge of the hydrogeologic factors, however, the only way to know for certain what is underground is to drill. Hence, the drilling (and testing) of exploratory wells is essential prior to the development of a production well.

Carpenter (1981, p. 68) states, "The value of an exploratory test hole prior to construction of a production well, especially when a high yield well is the objective, has often been proven to be well worth the added cost to the owner." Information obtained from drilling a test well includes

- expected well yield
- depth, thickness and type of water-bearing zones
- depth to non-pumping water level
- water temperature and quality

Figure 9.14 illustrates an example of a driller's log from a test well. With the test well data in hand the project engineer can determine the depth and diameter of casing, the specifications for the well screen, and other construction details.

In planning and carrying out a drilling program, several fundamental rules should be considered:

- Formulate working hypotheses concerning the most promising drilling sites.
- Drill the first hole at a site that promises early success.
- Regard exploration holes as an investment in the acquisition of knowledge, not as unproductive drilling.
- Analyze failures and draw the necessary conclusions.
- Substantiate all assumptions, especially disagreeable ones.
- Maintain continuous evaluation and feedback of all information acquired by drilling, whether successful or not.

All of the data and technical specifications necessary for drilling at a test

ROTARY WELL DRILLING, INC.
R.D.#1, BOX 555
HUMMEL'S WHARF, PA

MUNICIPAL WATER AUTHORITY
P.O. BOX 1000
RICHLAND, PA

TW #21   D R I L L   L O G

0" - 10'	BROWN ROCK
10' - 70'	TOUGH GREY ROCK, (SHALE)
70' - 98'	SANDSTONE AND SHALE LAYERS (GREY)
98' - 152'	GREY SANDSTONE, WITH QUARTZ AND SHALE LAYERS
152' - 153'	FRACTURED GREY SANDSTONE
153' - 171'	GREY SANDSTONE
171' - 172	FRACTURED GREY SANDSTONE
172' - 221'	GREY SANDSTONE WITH SHALE AND QUARTZ LAYERS
221' - 222'	FRACTURED, WATER
222' - 244'	GREY SANDSTONE WITH QUARTZ, SHALE AND PYRITE
244' - 247'	BROKEN FRACTURED GREEN AND GREY ROCK
247' - 273'	DARK GREY ROCK

WELL TOTALS:

273' OF 6" DRILLING	2 1/2 GPM AT 71'
21'5" OF 6 1/4" CASING	3 1/2 GPM AT 123'
1 6 1/4" DRIVE SHOE	1 GPM AT 148'
1 7 5/8" WELL CAP	16 GPM AT 152'
GAL PER MIN 115 TO 120	17 GPM AT 173' TO 198'
STATIC WATER LEVEL: 1'	38 GPM AT 222'
	37 GPM AT 241'

Figure 9.14 Example of a driller's log.

site are compiled and evaluated by a hydrogeologist. This information, and any hydrogeologic problems, are discussed thoroughly with the other technical personnel involved in the project, such as geologists, engineers, drillers, and equipment suppliers.

Driscoll (1986) presents an extensive and detailed discussion of drilling methods. These methods are appropriate also for the drilling of groundwater monitoring wells used to assess contamination problems.

9.4 DESIGNING GROUNDWATER MONITORING SYSTEMS

Groundwater monitoring projects fall into two broad categories: regional investigations, which seek to establish an overall picture of ambient water quality within all or parts of an aquifer, and site investigations, which are usually directed at assessing existing or potential contamination from one or more specific sources. In regional groundwater quality investigations, samples are collected routinely over a period of years so as to determine the changes in water quality over time. Moreover, these samples are commonly collected from existing public and private water supply wells, rather than from wells drilled specially for the purpose. In most site investigations, the goal is to determine the effect that a contaminant source has had, or may have in the future, on nearby groundwater quality. Special monitoring wells are sited and constructed for the purpose of defining the extent and geometry of the contaminant plume within and even outside of the boundaries of a particular facility. The analytical results of site investigations are especially important because the level of contaminant concentration may require specified regulatory action.

In general, the design and construction of monitoring wells follow the techniques of those of production wells; however, a monitoring well is built specifically to provide access to the groundwater so that a so-called "representative" sample of water can be withdrawn and analyzed. While well yield is important, the ability of the well to produce large quantities of water for supply purposes is not a primary objective. Instead, emphasis is placed on locating and constructing a well from which groundwater samples that will reveal reliable, meaningful water quality information are easily obtainable. Toward this end, the chemistry of suspected pollutants and the hydrogeologic setting play major roles in the drilling technique and the well construction materials used. It is particularly important that the materials and techniques used to construct the monitoring well do not alter the quality of the water being sampled.

Furthermore, it is important to recognize that potential pathways for contaminant migration are three dimensional. Typically, dissolved constituents will descend vertically through the unsaturated zone beneath the contaminated site and then, upon reaching the saturated zone, move horizontally in the direction of groundwater flow. Consequently, the design of a monitoring network that intercepts these potential pathways requires attention to both the horizontal placement of the monitoring wells and their depth.

Several components need to be considered in the design of a monitoring system, including location and number of wells, screen length and depth of placement, diameter, and casing and screen material.

In any site investigation the location and the number of monitoring wells are closely linked and are determined chiefly by the specific goals of the monitoring program. There is no universal rule for establishing the location and number of monitoring wells. The hydrogeologic setting of the site, the characteristics of the source contaminants, and the size of the area under investigation all help to determine where and how many wells should be constructed. For example, the more complicated the geology and hydrology, the more complex the contaminant source; and the larger the area being investigated, the greater the number of monitoring wells that will be required.

For waste sites that are regulated under the Resource Conservation and Recovery Act (RCRA), the minimum number of monitoring wells for a facility is four—one upgradient well and three downgradient wells (see Figures 9.15 and 9.16).

The purpose of an upgradient monitoring well is to provide background water quality data. Under RCRA, upgradient wells must be sited beyond the upgradient extent of potential contamination from the contaminant source so as to provide samples representative of background water quality. While the regulatory requirement for the minimum number of upgradient wells is one, there should be a sufficient number of wells to account for any heterogeneity in background groundwater quality.

The primary purpose of a downgradient monitoring well is to detect contaminants should they leak from the waste unit. In order to satisfy regulatory requirements for immediate detection of any contamination, downgradient monitoring wells must be located at the downgradient perimeter of the waste unit (see Figures 9.15 and 9.16). The precise place-

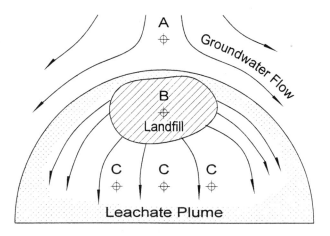

Figure 9.15 Map illustrating the location of monitoring wells at an RCRA site.

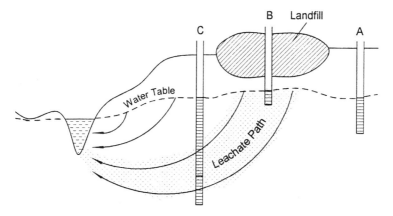

Figure 9.16 Cross section illustrating the locations of monitoring wells at an RCRA site.

ment of detection monitoring wells along the downgradient edge of the waste unit should be predicated on the interception of potential pathways for contaminant migration. The horizontal distance between wells should be based upon site-specific factors. For example, the interval between wells may be relatively wide if the site exhibits simple geologic setting, homogeneous hydraulic conductivity, low hydraulic gradient, and high dispersivity potential. If the site exhibits complex geology, heterogeneous hydraulic conductivity, high hydraulic gradient, low dispersivity potential, and is situated near or on a recharge zone, then the well interval should be relatively narrow. Additionally, care should be taken to site monitoring wells on discrete zones of potential contaminant migration, such as fracture traces or solution channels.

The third dimension to monitoring of existing or potential contaminant pathways is provided by the proper selection of the vertical sampling interval. Decisions regarding vertical sampling intervals are based on the results of a thorough site characterization, which defines both the depth and thickness of the stratigraphic zones that could serve as contaminant pathways.

In order to ensure comparability of data, upgradient and downgradient wells must be screened in the same stratigraphic zone(s). The depths at which downgradient wells must be screened are related to both the hydrogeologic factors influencing the pathways of contaminant migration and the chemical and physical characteristics of the contaminants. For example, the presence of "sinkers" or "floaters" (i.e., immiscible liquid phases) in the contaminant plume will affect decisions concerning screen length and depth. Furthermore, serious consideration must be given to natural fluctuations of the potentiometric surface, which can amount to several feet over a period of a year or more, and to artificial fluctuations caused largely by pumping, which can amount to several tens of feet in only a few hours.

A thorough review of groundwater monitoring techniques is presented by Nielson (1990).

9.5 REFERENCES

Campbell. M. D., and Lehr, J. H., 1973. *Water Well Technology.* New York, NY: McGraw-Hill Book Co.

Carpenter, C., 1981. The value of exploratory test holes in ground water development. *Water Well Journal,* November, pp. 68–69.

Caswell, W. B., 1979. *Ground Water Handbook for the State of Maine.* Maine Geological Survey.

Couch, J. V. 1982. *Fundamental Statistics for the Behavioral Sciences.* New York, NY: St. Martin's Press.

Davis, S. N., and DeWiest, R. J. M., 1966. *Hydrogeology.* New York, NY: John Wiley & Sons, Inc.

Driscoll, F. G., 1986. *Groundwater and Wells* (2nd ed.). St Paul, MN: Johnson Division.

Evans, R. B., 1982. Currently available geophysical methods for use in hazardous waste site investigations. In *Risk Assessment at Hazardous Waste Sites,* pp. 93–115. American Chemical Society.

Faill, R. T., and Wells, R. B., 1974. *Geology and Mineral Resources of the Millerstown Quadrangle, Perry, Juniata, and Snyder Counties, Pennsylvania.* Pennsylvania Geologic Survey Atlas 136.

Fetter, C. W., 1994. *Applied Hydrogeology.* 3rd ed. New York: Macmillan College Publ. Co.

Freeze, R. A., and Cherry, J. A., 1979. *Groundwater.* Englewood Cliffs, NJ: Prentice-Hall, Inc.

Gibson, U. P., and Singer, R. D., 1971. *Water Well Manual.* Berkeley, CA: Premier Press.

Johnston, H. E., 1970. *Ground-water Resources of the Mifflintown and Loysville Quadrangles in South-Central Pennsylvania.* Pennsylvania Geologic Survey Water Resource Report 27.

Lahee, F. H., 1961. *Field Geology.* Sixth edition. New York: McGraw-Hill Book Co.

Lattman, L. H., 1958. Technique of mapping geologic fracture traces and lineaments on aerial photographs. *Photogrammetric Engineering,* 84:568–576.

Lattman, L. H. and Parizek, R. R., 1964. Relationship between fracture traces and the occurrence of ground water in carbonate rocks. *J. of Hydrology,* 2:73–91.

Lennon, G. P., and Meyers, P. B., 1988. *Review of the Geological Conditions and the Hydrogeologic Impact of the Reactivation of PP&L Ash Basin #2, Sunbury SES, Monroe Township, Snyder County, Pennsylvania.* Lehigh University, IMBT Hydraulics Laboratory.

Meiser & Earl, Hydrogeologists, 1982. *Use of Fracture Traces in Water Well Location: A Handbook.* U.S. Dept. of the Interior.

Nielson, D. M. (ed.), 1990. *Practical Handbook of Ground Water Monitoring.* Chelsea, MI: Lewis Publ.

Siddiqui, S. H., and Parizek, R. R., 1971. Hydrogeologic factors influencing well yields in folded and faulted carbonate rocks in central Pennsylvania. *Water Resources Research,* 7(5):1295-1312.

CHAPTER 10

Case Study: Siting Test Wells for a Public Water Supply System

10.1 STATEMENT OF THE PROBLEM

A rapidly growing municipality is seeking to expand its public water supply in order to meet a usage of 500,000 gallons per day. Because of unfavorable hydrologic and engineering factors, surface water has been ruled out as a source of this additional supply; thus, the future supply must come from the groundwater of the region.

10.2 PURPOSE AND SCOPE

The purpose of the project is to identify one or more well sites from which approximately 500,000 gallons per day of water could be produced. The specific objectives of the project are to

- characterize the geology of the region
- determine the availability of groundwater from the aquifers
- locate the optimum sites for exploratory test wells
- evaluate the exploratory test well sites

10.3 BACKGROUND AND PROJECT SETTING

The municipality is situated in the physiographic Valley and Ridge region of central Pennsylvania. The water needs of the residents, commercial establishments, and industries of the municipality are supplied by a public water authority. Currently this water authority serves a population of approximately 500 persons with 20 million gallons of water annually. This water comes from two wells which are situated on the floodplain of a moderate-size stream, at the north edge of the community. The raw water is chlorinated and stored temporarily in a closed, concrete reservoir near the pump house. The treated water is distributed throughout the community through a 10-inch main and 6-inch distribution pipes.

Table 10.1 illustrates the stratigraphic sequence of the region. The formations are comprised of strata that range in age from early Silurian to late Devonian. As the geologic base map (Figure 11.1 in Chapter 11) shows, the formations form long and narrow outcrop bands which are characteristic of the eroded fold structures of this region.

10.4 METHODS OF INVESTIGATION

In order to identify the optimum sites for public water supply wells, the investigative methods described below will be employed:

(1) Compilation and review of literature and data sources of the hydrogeology of the project region, including published reports, maps, unpublished theses, aerial photographs, and driller's records
(2) Characterization of the regional geologic setting by the preparation of a regional geologic map, geologic column, structure section, and fracture trace map (Chapter 11)
(3) Analysis and interpretation of well yield data of the aquifers of the region, including the calculation and graphing of well-yield frequency distributions (Chapters 12 and 13, respectively)
(4) Preparation of a groundwater availability map (Chapter 14)

TABLE 10.1. Stratigraphic Sequence of the
Case Study Project Region.

Formation	Map Symbol	Average Thickness (ft.)
Catskill	Dciv, Dcsc	Top not exposed
Trimmers Rock	Dtr	2500
Hamilton	Dh	1200
Onondaga-Old Port	Doo	250
Keyser-Tonoloway	DSkt	700
Wills Creek	Swc	600
Bloomsburg-Mifflintown	Sbm	610
Clinton	Sc	100
Tuscarora	St	Base not exposed

TABLE 10.2. An Example of an Outline for a Groundwater Exploration Report.

Section	Products and Exhibits
TITLE PAGE	
1.0 INTRODUCTION	
1.1 Purpose and Scope	
1.2 Project Setting	Index map of project region
2.0 METHODS OF INVESTIGATION	
2.1 Collection of Background Data	
2.2 Data Analysis and Presentation	
2.3 Field Surveys	
3.0 RESULTS OF INVESTIGATION	
3.1 Geologic Setting	Geologic map
	Geologic column
	Structure section(s)
	Fracture trace map
3.2 Analysis of Well Yield	Well yield frequency table(s) and graphs
3.3 Groundwater Availability	Groundwater availability map(s)
3.4 Optimum Test Well Sites	Test well site map
4.0 CONCLUSIONS AND RECOMMENDATIONS	
5.0 REFERENCES	

(5) Identification and evaluation of one or more promising sites for drilling exploratory test wells, by means of geologic, hydrologic, and surface geophysical methods (Chapter 15)

10.5 REPORT OF FINDINGS

The results of the project can be summarized in a report (see Table 10.2 for an example of an outline).

CHAPTER 11

Method for Characterizing the Hydrogeologic Setting

11.1 INTRODUCTION TO METHODS

CHARACTERIZATION of the hydrogeologic setting of a region or a site is an important preliminary step in any groundwater investigation. As the information in Chapter 10 demonstrates, a complete characterization of hydrogeologic setting requires much detailed data and analysis, which is time-consuming and expensive. Considerable knowledge may be gained about the project region or site, however, by a careful examination of existing information.

The preliminary purpose of characterization is to ascertain the hydrogeologic conditions as they apply to locating test wells for water supply projects or monitoring wells for contamination. Knowledge of the petrographic, stratigraphic, and structural characteristics will enable an investigator to identify potential water-bearing formations, establish their geographic distribution and depths, and delineate zones of fracture concentrations.

This chapter describes the preparation of four geologic exhibits which are used in the characterization of a project region or site: a *geologic base map* (Section 11.2), a *geologic column* (Section 11.3), a *structure section* (Section 11.4), and a *fracture trace map* (Section 11.5).

11.2 PREPARING A GEOLOGIC BASE MAP

Purpose

To prepare a geologic base map for a groundwater investigation.

133

Reference

Lahee, F. H., 1952, *Field Geology.* New York: McGraw-Hill Book Co., pp. 691–697.

Equipment and Materials

- geologic base map (e.g., Figure 11.1)
- straight-edge ruler
- colored pencils
- drafting pencil (3H)

Procedure

A description of a geologic map and an example is presented in Chapter 9. More detailed information is given by Lahee (1952). If you are using the example problem in the case study described in Chapter 10, skip the first two steps below and go on to Add colors or patterns.

Construct a base map. Use the hand-drafting techniques described in Chapter 6 to prepare a base map of the project site or region. Alternatively, an available topographic or planimetric map may be employed as a base map for overlaying the geologic information.

Overlay the formational boundaries

(1) In light pencil trace the boundary lines of each of the formations.

(2) Draft the appropriate map symbol (e.g., Dc, Swc, etc.) for each formation. For an example, see Figure 11.1.

Add colors or patterns. Use colored pencils to shade each formation on the map (see, e.g., Figure 11.1) in a distinctive color or unique pattern (e.g., stippled, hachured, etc.) to distinguish the different formations). (It is helpful, though not essential, to employ colors or patterns that match those of the same formations on published geologic maps of the region.)

Add structural symbols. Draft the appropriate structural symbols to denote the traces of the axial surfaces of the folds.

Complete the map to report quality. For a hand-drafted map, carefully trace over all pencil lines in India ink and erase all residual pencil marks.

11.3 PREPARING A GEOLOGIC COLUMN

Purpose

To construct a geologic column representative of a project region.

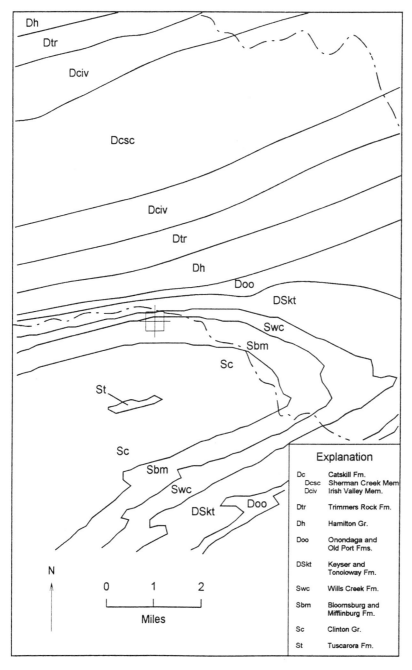

Dh
Dtr
Dciv
Dcsc
Dciv
Dtr
Dh
Doo
DSkt
Swc
Sbm
Sc
St
Sc
Sbm
Swc
DSkt
Doo

Explanation

Dc	Catskill Fm.
Dcsc	Sherman Creek Mem
Dciv	Irish Valley Mem.
Dtr	Trimmers Rock Fm.
Dh	Hamilton Gr.
Doo	Onondaga and Old Port Fms.
DSkt	Keyser and Tonoloway Fm.
Swc	Wills Creek Fm.
Sbm	Bloomsburg and Mifflinburg Fm.
Sc	Clinton Gr.
St	Tuscarora Fm.

N

0   1   2

Miles

Figure 11.1 Geologic base map of project region (based on Berg and Dodge, 1981, p. 229).

135

Reference

Lahee, F. H., 1952. *Field Geology.* New York, NY: McGraw-Hill Book
   Co., pp. 689–690.

Equipment and Materials

- formation thickness data (e.g., Table 10.1)
- electronic calculator
- geologic column form (e.g., Figure 11.2) or arithmetic graph
  paper (e.g., K & E 460780)
- ruler/scale
- colored pencils
- drafting pencil (3H)

Procedure

A description of a geologic column and an example is presented in
Chapter 9. More detailed information is given by Lahee (1952).
   Calculate the total thickness of the stratigraphic sequence.

(1) Sum the thicknesses of the individual formations of the sequence
    under investigation (for the case study in Chapter 10, see Table 10.1).
(2) Record this value on the line below.
    Total thickness:_____ feet or meters

   Establish a relative scale for the geologic column.

(1) Divide the total thickness of the sequence (in feet or meters) by the
    height of the geologic column (in inches or centimeters) illustrated in
    Figure 11.2.
(2) In order to make drafting of the geologic column easier, round off the
    value of the relative scale to the next largest even hundred units.
(3) Record this value on the line below.
    Scale: 1 inch/centimeter = _____ feet/meters

   Fix the position of the base of the youngest formation.

(1) Measure down about one-half inch from the title base line to the
    geologic column.
(2) Draw a horizontal line to mark the base of the youngest formation
    (e.g., Dck) across the width of the entire column form.

   Establish the relative positions of the formational boundaries.

(1) Using the relative scale (feet/inch or meters/centimeters), measure

Thick-ness feet	Description	Age	Formation	Map Sym-bol	Graphic Column
-					
-					
-					
-					
-					
-					
-					
-					
-					
-					
-					
-					
-					
-					
-					
-					
-					
-					
-					
-					
-					
-					
-					
-					

Figure 11.2 Geologic column form.

137

down from the base of the youngest formation a distance in inches or centimeters that represents the thickness of the formation directly beneath the youngest formation.

(2) Draw a horizontal line, which represents the base of this formation, across the geologic column.

(3) Repeat this procedure until the stratigraphic boundaries of all formations have been marked off on the column.

Add descriptive information for the formations. In each of the appropriate columns print 1) the numerical value of the thickness, 2) a brief description of the rock type, 3) the name of the formation, and 4) the map symbol (e.g., Dh).

Add geologic age information. Mark the boundaries of each geologic period (e.g., Devonian) in the Age column by drawing a horizontal line across the entire width of the geologic column and printing in the names of each of the periods.

Construct a graphic log. Shade each of the formations in the graphic column in a color that matches its geologic map color (see Figure 11.1), or use unique standard patterns to distinguish the formations.

11.4 PREPARING A STRUCTURE SECTION

Purpose

To construct a structure section representative of a project region.

Reference

Lahee, F. H., 1952. *Field Geology.* New York, NY: McGraw-Hill Book Co., pp. 630–637.

Equipment and Materials

- geologic base map (e.g., Figure 11.1)
- arithmetic graph paper (e.g., Figure 11.2 or K & E 460780)
- drafting triangle (useful but not essential)
- colored pencils
- drafting pencil (3H)

Procedure

A description of a structure section and an example is presented in Chapter 9. More detailed information is given by Lahee (1952).

Prepare a profile base line.

(1) Scan the base map and determine the major structural trend of the region (i.e., roughly northwest to southeast in Figure 11.1).

(2) Beginning along one edge of the map, construct a base line perpendicular to the major structural trend across the map to the opposite edge.

Mark the positions of the formational boundaries.

(1) Begin the construction of the structure section by laying the top line of the grid form (e.g., Figure 11.3) along the profile base line of the map and align it so that the left edge of the grid lies at one border line of the map (for the base map of the case study, use the north border).

(2) Beginning at this border of the map, scan along the top of the grid form to the first formational boundary and mark its location on the top line of the grid with a pencil.

(3) Move to the next formational boundary and mark its position.

(4) Between the two marks write in the map symbol for the formation that occupies that outcrop width.

(5) Repeat the procedure across the profile line until the location of all formational boundaries have been marked and labeled.

Determine and mark dip directions.

(1) Using the relative ages of the formations as a guide, determine the dip direction of the strata at each formational boundary. (Except in regions where the strata are overturned, dip is in the direction of the younger rocks of the sequence.)

(2) Use a short line to mark the dip direction at the formational boundary (at an angle of 45° for the project in the case study).

(3) Repeat this procedure until the dip directions of all of the formations have been marked on the grid. Be careful to recognize any reversals in dip direction that occur where the base line crosses the traces of fold axes.

Construct the subsurface configuration of the formations. Keeping formational boundaries parallel with one another (i.e., the formations should retain constant thickness everywhere), sketch in the subsurface configuration of the formations. For an example, see Figure 9.3.

Add labels and colors.

(1) Complete the structure section by labeling each formation with its map symbol and shading it to match its map color (see Figure 11.1).

(2) For production quality exhibits, carefully trace over the pencil lines and labels in India ink and erase all extraneous pencil marks.

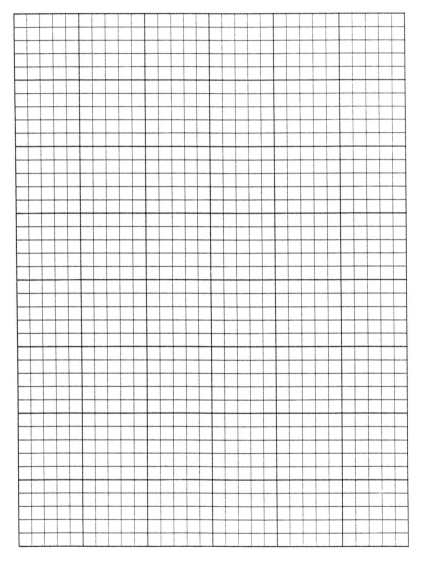

Figure 11.3 Structure section grid form.

140

11.5 PREPARING A FRACTURE TRACE MAP

Purpose

To prepare a fracture trace map of a project region.

References

Lattman, L. H., 1958. Technique of mapping geologic fracture traces and lineaments on aerial photographs. *Photogrammetric Engineering,* 84:568–576.

Meiser & Earl, Hydrogeologists, 1982. *Use of Fracture Traces in Water Well Location: A Handbook.* U.S. Dept. of the Interior.

Equipment and Materials

- aerial photographs of the project region (stereo pairs)
- base map of site or region (e.g., Figure 11.4)
- pocket or mirror stereoscope
- ruler/scale
- soft yellow pencil (F)

Procedure

A description of a fracture trace map and an example is presented in Chapter 9. The procedure described here assumes that the investigator has some experience in the identification of fracture traces on aerial photographs. In any event, Lattman (1958) and Meiser & Earl (1982) provide useful information about this technique.

Position the aerial photographs for interpretation.

(1) Scan the aerial photograph(s) of the project region in order to obtain a sense of the general orientation or structural pattern exhibited by fracture traces or other linear features.

(2) Orient a stereo pair of photos on a table so that the shadows "fall" toward you and select a distinctive feature to focus on.

Position the stereoscope.

(1) Align the stereoscope over the photos and adjust it so that the center of the lenses are the same distance apart as the pupils of your eyes.

(2) Focus the lenses, if necessary.

Focus on a selected feature.

(1) Lift and move the stereoscope over the photos until a distinctive

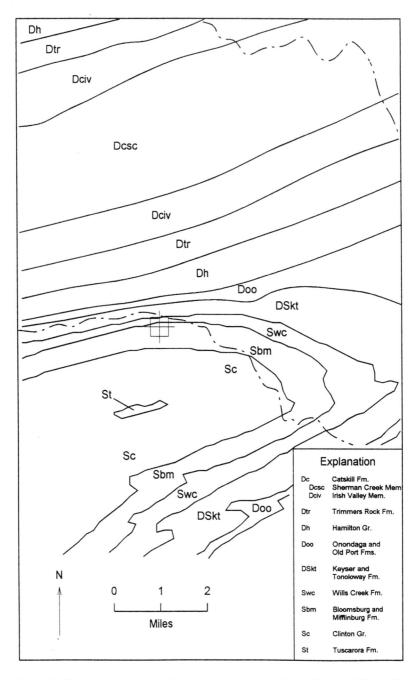

Figure 11.4 Fracture trace base map of project region (based on Berg and Dodge, 1981, p. 229).

142

feature (large structure, pond, etc.) lies under the centers of the two lenses (or mirrors).

(2) Then, shift or rotate one of the photos until a three-dimensional image appears. (If you are new at this technique, be patient. The procedure sometimes takes a little practice.)

Identify and mark linear features.

(1) After the photos have been aligned properly, move the stereoscope over the stereo pair and mark the ends of the linear features observed (soil tone, vegetation alignment, straight stream reach, and valley alignment, etc.) with a soft yellow pencil on one of the photos. The ends of each line should be marked by arrows approximately 3 mm long, with barbs of the arrows pointing toward one another. *Warning!* Do not use a ballpoint pen or a hard pencil. They would leave a permanent mark or groove in the photo.

(2) Repeat this procedure until all linear features have been marked on the photo. (*Note:* All linear features will not necessarily denote fracture traces, but all should be mapped.)

Transfer the linears to a base map. Replicating accurately the length and azimuth of each linear feature on the aerial photograph, draw a straight line to mark its location on the fracture trace base map (e.g., Figure 11.4).

Complete map to production quality. Trace over the penciled linears in India ink and erase any residual pencil marks.

11.6 REFERENCES

Berg, T. M. and Dodge, C. M., 1981. *Atlas of Preliminary Geologic Quadrangle Maps of Pennsylvania.* Pennsylvania Geological Survey Map 61.

Lahee, F. H., 1952. *Field Geology.* New York, NY: McGraw-Hill Book Co.

Lattman, L. H., 1958. Technique of mapping geologic fracture traces and lineaments on aerial photographs. *Photogrammetric Engineering,* 84:568–576.

Meiser & Earl, Hydrogeologists, 1982. *Use of Fracture Traces in Water Well Location: A Handbook.* U.S. Dept. of the Interior.

CHAPTER 12

Analytical Procedure for Determining Well Yield Frequency Distributions

12.1 INTRODUCTION TO METHODS

A list of reported well yields from an aquifer of a region (see Table 12.1 for example) provides some guidance for determining how much groundwater one might expect to obtain from wells drilled into the aquifer. A simple list of well yields, however, is not organized in a way that clearly reveals groundwater availability. Statistical measures such as range, central tendency, or variation are not clearly evident from such a list.

An estimation of the availability of groundwater in an aquifer – and, thus, a comparison of water yield from different aquifers of a region – begins with the construction of a frequency distribution of reported well yields. As described in Chapter 9, a frequency distribution table displays how many wells of the aquifer yield a given amount of water. These data, then, may be used to estimate the chances (e.g., 10 wells out of 100) of obtaining a specified yield (e.g., 15 gpm) from a well drilled randomly into the aquifer.

This chapter describes two methods which can be used to create frequency distributions of well yield: the *Electronic Calculator and Worksheet Method* (Section 12.2) and the *Microcomputer and Electronic Spreadsheet Method* (Section 12.3). Table 12.1 displays a list of reported well yields from the case study in Chapter 10. These data (or your own project data) may be used for the application of the methods described in this chapter. [For additional information, see Chow (1964) and Lapin (1990) or other introductory statistics textbooks.]

145

TABLE 12.1. Reported Well Yields
of the Onondaga-Old Port Formation.

Well No.	Well Yield (gpm)	Well No.	Well Yield (gpm)
CO-051	185	MT-132	3
CO-188	12	MT-140	30
CO-307	88	MT-142	5
CO-312	9	MT-143	20
JU-118	8	MT-144	6
JU-183	2	MT-190	10
MF-056	5	MT-202	28
MF-093	10	MT-203	120
MF-107	20	MT-232	6
MF-109	20	NU-018	18
MF-184	10	NU-178	20
MF-186	4	PE-078	44
MF-200	10	PE-143	25
MF-201	30	PE-196	12
MF-219	10	PE-225	4
MT-016	158	PE-233	20
MT-017	122	PE-390	6
MT-032	73	PE-391	9
MT-048	5	PE-533	3
MT-065	20	PE-552	35
MT-066	50	PE-564	20
MT-067	20	PE-624	60
MT-068	12	PE-630	12
MT-079	12	PE-675	20
MT-081	8	SN-003	8
MT-082	3	SN-097	10
MT-083	60	SN-149	40
MT-086	50	SN-211	20
MT-089	3	SN-224	15
MT-090	15	SN-283	25
MT-091	30	SN-314	5
MT-125	40		

12.2 ELECTRONIC CALCULATOR AND WORKSHEET METHOD

Purpose

To create a frequency distribution of well yields.

Reference

Lapin, L. L., 1990. *Statistics for Modern Business Decisions* (5th ed.). New York, NY: Harcourt Brace Jovanovich Publ., pp. 11–13.

Equipment and Materials

- reported well yield data (e.g., Table 12.1)
- electronic calculator
- well yield frequency worksheet (e.g., Figure 12.1)
- pencil (3H)

Procedure

Identify and mark the well yields in each class interval. Read down a list of reported well yields (e.g., Table 12.1) well by well. Make a tally mark (/) in column 2 of the well yield frequency worksheet (Figure 12.1) beside the appropriate class interval that contains the well yield value of each well. (For the case study data on Table 12.1, begin by making a tally mark beside the > 101 class interval for Well CO-051, another beside 11–20 class interval for Well CO-188, and so on until all wells have been tallied.)

Determine the number of well yield values in each class interval.

(1) Count the number of tally marks (i.e., the frequency) within each of the class intervals.

(2) Record the result in column 2 under f.

Calculate the total number of well yield values in the sample population.

(1) Sum the frequencies (f) in column 2.

(2) Record the value at the bottom of the column next to $\Sigma f = \_\_$.

Compute the relative frequency (rf) for each class interval.

(1) Divide the frequency of each class interval by the total number of wells (i.e., $f/\Sigma f$).

(2) Record the results in column 3.

Compute the relative cumulative frequency (rcf) for each class interval.

1	2		3	4	5
Well Yield gpm	Number of Wells tally marks	f	Relative Frequency rf	Relative Cumulative Frequency rcf	Percent-ile Rank = or >
≤2					
3-5					
6-10					
11-20					
21-30					
31-40					
41-50					
51-60					
61-70					
71-80					
81-90					
91-100					
≥101					
	Σf =				

Figure 12.1 Well yield frequency worksheet.

(1) Add the value of the relative frequency (rf) of the first class interval to that of the next (i.e., second) class interval. Record the sum in the column 4 table box on the second class interval row.

(2) Add this sum to the relative frequency value of the third class interval. Record the result in the column 4 table box on the third class interval row.

(3) Repeat this procedure until values of cumulative relative frequency have been calculated for all of the class intervals.

Compute the percentile rank (= or >) of each class interval.

(1) Multiply the value of the relative cumulative frequency in each class interval by 100 (i.e., rcf × 100).

(2) Subtract this result from 100.

(3) Record the precentile results in column 5.

12.3 MICROCOMPUTER AND ELECTRONIC SPREADSHEET METHOD

Purpose

To calculate a frequency distribution of well yields.

Equipment and Materials
- reported well yield data (e.g., Table 12.1)
- microcomputer with an 80386 processor or higher, MS-DOS 3.3 or higher, Windows 3.1 or higher, hard drive, floppy disk drive, and graphics printer
- Windows version of electronic spreadsheet software (e.g., Lotus 1-2-3, Quattro Pro, or Excel)
- Hydrodata Diskette

Procedure

This method employs electronic spreadsheet software to calculate a well yield frequency distribution. First, values of reported well yield (use Table 12.1 or your own project data) are entered onto a blank worksheet. Then, these data are combined with the spreadsheet calculation template (filename: yield.txt), which is loaded from the Hydrodata Diskette. The frequency calculations are accomplished by means of formulas entered into the cells of the spreadsheet template. When the calculations are completed, a table of the data may be printed.

Starting Up

Turn on the computer, monitor and printer. Wait until the Windows desktop is displayed.
Open the application.

(1) Open the Windows group icon that contains your software application (e.g., Quattro Pro), or click on the Windows 95 Start button and point first to Programs and then to the application folder.
(2) Double click on the application icon from the group window, or click on the name of the application from the drop-down list.
(3) Insert the Hydrodata Diskette into drive A (or B).

Entering Well Yield Values

The first eight rows of the worksheet should be left blank in order to accommodate a template file which will be loaded and combined later.

Enter the first well code.

(1) Select cell A9.

(2) Type the well code designation (e.g., CO-051).

Enter the first well yield value.

(1) Select cell B9.

(2) Type the value of the well yield (e.g., 185).

Complete the well yield data file. Repeat the data entry procedure down the spreadsheet until all well codes and well yield values have been entered into columns A and B.

Loading and Overlaying the Data Distribution Template

Quattro Pro

(1) Choose Notebook | Combine.

(2) Click on the arrow next to the File Type text box.

(3) Select Text file type.

(4) Select the well yield frequency template file from the drop-down list (e.g., yield.txt).

(5) Click on OK. The well yield frequency template will be combined with the data file and displayed on the screen.

1-2-3

(1) Choose File | Open. The Open File dialog box will be displayed.

(2) Click on the arrow next to the File Type text box.

(3) Select Text file type.

(4) In the File Name drop-down box, select the data distribution template file (e.g., yield.txt).

(5) Click on Combine. The well yield frequency template will be combined with the data file and displayed on the screen.

Excel

(1) Select the well data block (e.g., cells A9-B71).

(2) Choose Edit | Copy. The well yield data will be copied to Clipboard.

(3) Choose File | Open.

(4) Click on the arrow next to the File Type text box.

(5) Select Text file type.

(6) Select the file name of the data distribution template (e.g., yield.txt).

(7) Click on OK. The Excel Text Import Wizard will appear.

(8) Click twice on Next> and once on Finish>. The well yield frequency template will open on a new worksheet.

(9) Select cell insert point (e.g., cell A9).

(10) Choose Edit|Paste. The well data block will be copied into the well yield frequency template.

Entering Spreadsheet Formulas

For Microsoft Excel users, be sure to enter an equal sign (=) in front of all the formulas given below.

Enter calculation formulas.

(1) Select cell F7.

(2) Type (E7/E$20).

(3) Select cell G7.

(4) Type (F7).

(5) Select cell G8.

(6) Type (F8 + G7).

(7) Select cell H7.

(8) Type ((100)-G7*100).

(9) Select cell E20.

(10) Type @SUM(E7..E19).

Copy calculation formulas.

(1) Select cell F7.

(2) Choose Edit|Copy.

(3) Select cells F8–F19.

(4) Choose Edit|Paste.

(5) Select cell G8.

(6) Choose Edit|Copy.

(7) Select cells G9–G19.

(8) Choose Edit|Paste.

(9) Select cell H7.

(10) Choose Edit|Copy.

(11) Select cells H8–H19.

(12) Choose Edit|Paste.

(13) Select cell E20.

(14) Choose Edit|Copy.

(15) Select cell F20.

(16) Choose Edit|Paste.

Creating a Data Distribution Table

Quattro Pro

(1) Select column D.

(2) Choose Block | Delete.

(3) Choose Tools | Numeric Tools | Frequency. The Frequency Tables dialog box will be displayed.

(4) Click on the arrow next to the Value Block edit line.

(5) On the worksheet, select (i.e., drag and click on) the range of well yield values (e.g., B9–B71).

(6) Click on the arrow on the Frequency Tables menu bar to return to the worksheet. The selected cell range will appear on the Value Block edit line.

(7) Click on the arrow next to the Bin Box edit line.

(8) Select the range of cells that indicate the class intervals (e.g., C7–C18).

(9) Click on the arrow next to the Frequency Tables menu bar to return to the worksheet. The selected cell range will appear on the Bin Block edit line.

(10) Click on OK. The computed frequencies will be calculated and displayed in columns D–G.

1-2-3

(1) Select column D.

(2) Choose Edit | Delete.

(3) Choose Range | Analyze | Distribution.

(4) On the worksheet, select the range of well yield values (e.g., B9–B71). This cell range will appear on the Range of Value edit line.

(5) Click on the arrow next to the Bin Range line.

(6) Select the range of cells that represent the bin (e.g., C7–C18). The cell range will appear on the Bin Box edit line of the dialog box.

(7) Click on OK. The computed frequencies will be calculated and displayed in columns D–G.

Excel

(1) Choose Tools | Data Analysis. The Data Analysis dialog box will be displayed.

(2) In the Analysis Tools drop-down box, select Histogram.

(3) Click on OK. The Histogram text box will be displayed.

(4) In the Input text box, select the Input Range edit line. In the next few

steps, you may have to shift the dialog box and/or the worksheet in order to select the correct cells.

(5) Select the range of well yield data (e.g., cells B9–B71).

(6) Select the Bin Range edit line.

(7) On the worksheet select the bin range block (e.g., cells C7–C18).

(8) In the Output Options text box, select the Output Range edit line.

(9) Select an output range block starting cell (e.g., D6).

(10) Click on OK. A Microsoft Excel message box will appear stating that the output range will overwrite existing data. This action is expected and will cause no harm.

(11) Click on OK. The frequency output will appear on the worksheet in columns D and E, and new calculations will appear in columns F–H.

Modifying the Worksheet Format

(1) Select the Relative Frequency column of data and use the application formatting operations to change the numeric format to four decimals. (If necessary refer to the Help menus or the application user's manual.)

(2) Select the Cumulative Relative Frequency column of data and change the numeric format to two decimals.

(3) Select Recent Rank column of data and change the numeric format to one decimal.

Saving the Distribution Table File

Save the new file for the first time.

(1) Select cell A1.

(2) Choose File | Save As. The Save File dialog box will appear on the screen. The name of an original file (e.g., yield.txt) will be displayed and highlighted in the File Name text box.

(3) Press BACKSPACE to erase the name of the template file.

(4) Type the new frequency distribution file name (e.g., disdoo) in the File Name text box.

(5) Click on the arrow next to the File Type list box and select the type of file used by your particular application. (Do not use *.txt.)

(6) Click on OK. The new file will be saved to the drive and directory you specified. Later, to save the existing file choose File | Save. The changes to the file will be saved under the original file name and the old data will be overwritten.

Printing the Frequency Distribution Table

Before you print, make sure that your application is properly configured for your particular printer. The instructions below provide only basic information about printing the table. For more detailed instructions, see the application reference manual or click on the Help icon.

Check the printer configuration.

(1) Choose File│Print. The Print dialog box will appear on the screen.

(2) Make sure that your printer is identified as active. (If not, use the printer setup option to select and configure it.)

Edit page settings.

(1) Select the Distribution Calculations table block (e.g., cells C1–H20). (There is no need to print the entire list of well yield data.)

(2) Click on the Page Setup button, and, depending on your application, select:

Quattro Pro: Print to fit
1-2-3 in Size text box: Fit all to page
Excel in Scaling box: Fit to 1 page

(3) Click on OK.

Preview the print job.

(1) Click on the Print Preview button. A full-page view of the worksheet will be displayed. To view details, use the zoom option.

(2) Press ESC until the screen returns to the worksheet.

Print the table.

(1) Choose File│Print. The Print dialog box will be displayed. The selected block appears in the Print or Print What box.

(2) Click on Print (or OK). The well yield distribution table can now be printed. If the format of the printout requires modifying, or if an error in data or labels is evident, return to the worksheet and make the necessary corrections. (The application user's guide or reference manual may be useful for these procedures.)

Closing the Spreadsheet Application and Quitting Windows

Before closing the application, be sure to save all active files.

(1) Choose File│Exit. If you have saved all active files, the application window closes. If you changed an active file but did not save it, then the Exit dialog box will appear.

(2) If the Exit dialog box appears, choose Yes to save the file. The Windows desktop will appear on the screen.

(3) Open another application or quit Windows and turn off the computer. Be sure to close all open applications before quitting Windows.

12.4 REFERENCES

Chow, V. T. (ed.), 1964. *Handbook of Applied Hydrology.* New York, NY: McGraw-Hill Book Co.

Lapin, L. L., 1990. *Statistics for Modern Business Decisions* (5th ed.). New York, NY: Harcourt Brace Jovanovich Publ.

CHAPTER 13

Method for Graphing
Well Yield Distributions

13.1 INTRODUCTION TO METHODS

FREQUENCY distributions of well yields can be displayed effectively on lognormal probability graphs. See Chapter 9 for a brief description and an example of a lognormal plot. These graphs may be used to estimate measures of central tendency and dispersion of the distribution.

This chapter describes two methods which can be employed to construct plots of well yield frequency distribution graphs: the *Hand Plotting and Graphing Method* (Section 13.2) and the *Microcomputer and Graphing Software Method* (Section 13.3). A frequency distribution of the well yields of the Onondaga-Old Port Formation, which was calculated in Chapter 12, or your own project data, may be used for the application of the methods described in this chapter.

13.2 HAND PLOTTING AND GRAPHING METHOD

Purpose

To construct well yield frequency distribution graphs.

Equipment and Materials

- a frequency distribution of well yield data (see Chapter 12)
- lognormal graph paper (e.g., Figure 13.1 or K & E 468040)
- ruler/scale
- pencil (3H)

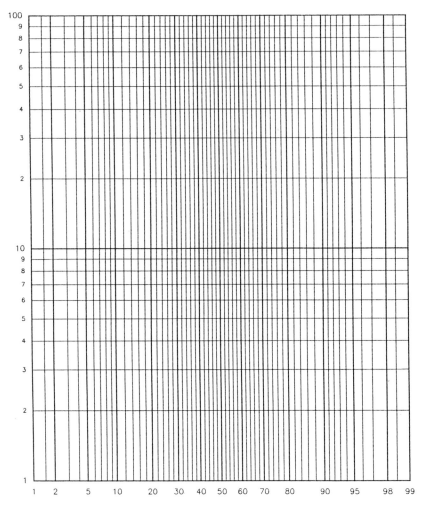

Figure 13.1 Lognornal probability graph form.

Procedure

Before continuing with this method, you should review the section in Chapter 9 on well yield frequency graphs.

Align the graph form. Orient a sheet of lognormal probability graph paper (e.g., Figure 13.1) so that the probability axis lies horizontally across the table in front of you. If you are using the graph form in Figure 13.1, skip the next two subsections and go on to Plot the data points.

Establish a scale for Y-axis. Lay out a well yield scale on the logarithmic axis (left vertical aixs). With one (i.e., 1) at the bottom of the axis, establish a log axis scale (e.g., 1, 10, 100, etc.) such that the range of well yields is accommodated by the necessary number of log cycles (see Figure 9.7, for example).

Establish a scale for the X-axis. Lay out a percentage frequency scale on the probability (bottom horizontal) axis and establish scale divisions that range from 5% at the left edge of the axis to 95% at the right edge (see Figure 9.8, for example).

Plot the data points. Plot each well yield–percentile rank data pair as a point on the graph. If you are using grouped data (e.g., Figure 12.1), select the maximum value of the well yield in each class interval for plotting on the Y-axis.

Construct an estimated "best-fit" line. Complete the graph by laying a straight-edge ruler down on the cluster of data points and drawing a straight line through the data points so that the line bisects the scatter of points.

Label the graph axes. Label the Y-axis "WELL YIELD (gpm)." Label the X-axis "PERCENT OF WELLS IN WHICH INDICATED YIELD WAS EQUALED OR EXCEEDED."

13.3 MICROCOMPUTER AND GRAPHING SOFTWARE METHOD

Purpose

To construct and display well yield frequency distribution graphs.

Equipment and Materials

- frequency distribution of well yield data (see Chapter 12)
- microcomputer with an 80386 processor or higher, MS-DOS 3.3 or higher, Windows 3.1 or higher, hard drive, floppy disk drive, and graphics printer
- GRAPHER for Windows (Version 1.25)
- Hydrodata Diskette

Procedure

This method employs GRAPHER for Windows to construct and display lognormal graphs of well yield frequency distributions. The GRAPHER worksheet can be used to create and save a data file of well yield values (use the results of the exercise in Chapter 12 for example or your own project data).

Starting Up

Turn on the computer, monitor and printer/plotter. Wait until the Windows desktop is displayed.

Open the application.

(1) Open the Windows group icon that contains your software application (e.g., Golden Software), or click on the Windows 95 Start button and point first to Programs and then to the application folder.

(2) Double click on the application icon (e.g., GRAPHER) from the group window, or click on the name of the application from the drop-down list.

(3) Insert the Hydrodata Diskette into drive A (or B).

Display the GRAPHER worksheet.

(1) Choose File|Worksheet. A worksheet window (labeled Sheet1) will appear superimposed over the plot window.

Entering Well Yield Distribution Data.

The well yield–percentile rank data will be arranged in two columns of the worksheet. Column A should contain the values of percentile rank (= or >) from a frequency distribution table, and Column B should contain the corresponding values of well yield (i.e., the maximum value of each class interval).

Enter the first percentile rank.

(1) Select cell A1.

(2) Type the rank (e.g., 98.4).

Enter the first well yield (i.e., class interval maximum) value, in gpm.

(1) Select cell B1.

(2) Type the well yield (e.g., 2).

Enter the second percentile rank.

(1) Select cell A2.

(2) Type the rank (e.g., 82.5).

Enter the second well yield value.

(1) Select cell B2.

(2) Type the well yield (e.g., 5).

Complete the data entry. Repeat the data entry procedure until all data pairs have been typed. After all the well yield distribution data have been entered, compare the data on the screen with the original data and correct all errors. When the data file is completed, save it to your data disk.

Saving the Well Yield Distribution Data File

Save the new file for the first time.

(1) Select cell A1.
(2) Choose File | Save As. The Save As dialog box will appear on the screen.
(3) Press BACKSPACE to delete any displayed text and type in a new file name (e.g., doofreq).
(4) Click on the arrow next to the File Type list box and select ASCII files [*.DAT].
(5) Click on the arrow next to the Drives list box and select the drive that contains the Hydrodata Diskette (e.g., A or B).
(6) Click on OK. The new file (e.g., doofreq.dat) will be saved to the drive and directory you specified. Later, to save the existing file choose File | Save.

Creating a Well Yield Frequency Distribution Graph

(1) Choose Window | Plot1.
(2) Choose File | Page Layout.
(3) In the Orientation text box, select Landscape.
(4) Click on OK.
(5) Choose Graph | Line or Symbol. The Select Worksheet dialog box will be displayed.
(6) Select the well yield data file (e.g., doo.dat).
(7) Click on OK. The Line Plot dialog box will be displayed. This box contains several specifications for the graph. (For more information, refer to the application user's manual.)
(8) Click on OK. The dialog box will disappear and an initial version of the graph will be displayed.

Editing the Graph Format

Edit the X-axis.

(1) Select (i.e., click on) the X-axis of the graph.

(2) Choose Set|Axis. The Edit X Axis dialog box will be displayed.

(3) Click on the arrow next to the Scale text box and select Probability.

(4) In the Length and Starting Position text box, click on the arrow next to the Length: text box until the value reads 5.80 in.

(5) Click on the arrow next to the X: text box until it reads 2.60 in.

(6) Click on the arrow next to the Y: text box until it reads 1.20 in.

(7) In the Axis Limits text box, select the Axis Min. text box, delete the default value (e.g., 0) and type the new value (e.g., 1).

(8) Select the Axis Max. text box, delete the default value (e.g., 100) and type the new value (e.g., 99).

(9) Choose the Edit Labels button. The Tick Labels dialog box will be displayed.

(10) Choose the Format button. The Label Format dialog box will appear.

(11) Click on the arrow next to the Decimal Digits text box until a zero (0) is displayed.

(12) Click on the OK button of each dialog box until the plot window is displayed again.

Edit the Y-axis

(1) Select the Y-axis of the graph.

(2) Choose Set|Axis. The Edit Y Axis dialog box will be displayed.

(3) Click on the arrow next to the Scale text box and select Log (base 10).

(4) In the Length and Starting Position text box, click on the arrow next to the Length: edit line until it reads 6.00 in.

(5) In the Length and Starting Position text box, click on the arrow next to the X: text box until the values reads 2.60 in.

(6) Click on the arrow next to the Y: text box until it reads 1.20 in.

(7) Select the Title text box.

(8) Type in the title of the Y-axis [e.g., WELL YIELD (gpm)].

(9) Choose the Edit Labels button. The Tick Labels dialog box will be displayed.

(10) In the Major Tick Labels text box, choose the Format button. The Label Format dialog box will appear.

(11) Click on the arrow next to the Decimal Digits text box until a zero (0) is displayed.

(12) Click on OK.

(13) In the Minor Tick Labels box, select Show Labels.

(14) Choose the Format button.

(15) Click on the arrow next to the Decimal Digits text box until zero (0) is displayed.

(16) Choose the Font button.

(17) Click on the arrow next to the Face text box and select the GS DEFAULT font style from the drop-down list.

(18) Click on the arrow next to the Points (i.e., font size) text box until 10 is displayed.

(19) Click on the OK button of each dialog box until the plot window is displayed again.

(20) Select View | Zoom Page. This action will resize the lognormal graph so that the entire page is in view.

Edit the Line.

(1) Choose Set | Line Attributes. The Line Attributes dialog box will be displayed.

(2) In the Width edit box, click on the arrow at the right until the value 0.010 is displayed.

(3) Click on OK.

Editing Grid Lines

(1) Select the X-axis.

(2) Choose Set | Grid Lines. The Grid Lines dialog box will be displayed.

(3) Select the At Major Ticks box.

(4) Click on OK.

(5) Select the Y-axis.

(6) Choose Set | Grid Lines. The Grid Lines dialog box will be displayed.

(7) Select both the At Major Ticks and the At Minor Ticks boxes.

(8) Click on OK.

Adding a Graph Title

(1) Choose Draw | Text. The normal cursor arrow will change to an arrow with a T.

(2) Position the cursor somewhere near the page coordinates X = 4.5, Y = 7.5. (The exact location is not necessary because the graph title may be repositioned at a later time.)

(3) Click on the chosen location. The Text dialog box will be displayed.

(4) Click on the arrow next to the Points edit line until the default number (e.g., 12) is replaced by a new font size (e.g., 16).

(5) In the text box at the bottom of the dialog box, type in the graph title (e.g., ONONDAGA-OLD PORT FM) in the space following the blinking cursor bar.

(6) Click on OK.

(7) On the graph, select (i.e., click on) the graph title.

(8) Click and drag the title to a suitable location in the left margin of the graph.

(9) Unselect the graph title by clicking once outside of the plot frame.

Saving the Well Yield Frequency Graph

Save the graph/chart and data.

(1) Choose File | Save As. The Save As dialog box will be displayed.

(2) With the default file name (e.g., plot1.grf) highlighted, type in the new name for the graph (e.g., doofreq.gr1).

(3) Click on OK. The graph will be saved under the specified file name.

Printing the Well Yield Frequency Graph

Before you print/plot, make sure that your application is properly configured for your particular printer. The instructions below provide only basic information about printing/plotting the graph. For more detailed instructions, see the application reference manual or click on the Help icon.

Check the printer/plotter configuration (optional).

(1) Choose File | Change Printer. The Change Printer dialog box will appear on the screen.

(2) Select your printer from the Printer list box. (Use Setup to specify print settings.)

(3) Click on OK.

Print/plot the graph.

(1) Make sure that the graph/chart window (not the worksheet) is displayed and active.

(2) Choose File | Print.

(3) Click on OK.

Creating Additional Lognormal Probability Graphs (optional)

The graph you have just created and saved can be employed as a template for plotting additional lognormal probability graphs of well yield data. Data files representing the well yield distribution of all of the aquifers of the case study (see Chapter 10) are stored on the Hydrodata Diskette. Each of these data files can be imported into the lognormal probability graph format already created for the Onondaga–Old Port aquifer. Follow the instructions below.

Display the graph template. If you do not have the lognormal probability graph of the Onondaga–Old Port aquifer opened and displayed in a GRAPHER Plot window, do so now.

Change the data file and plot a new graph.

(1) Select the line plot on the graph.

(2) Double click anywhere within the selection rectangle. The Line Plot dialog box will appear.

(3) Choose Change File. The Select Worksheet dialog box will appear.

(4) Select Open New Worksheet.

(5) Click on OK.

(6) Click on the arrow next to the List Files of Type drop-down box.

(7) Select ASCII Data (*.TXT). A list of the aquifer data files will appear in the File Name box. The aquifer codes are

dc	Catskill	sbm	Bloomsburg–Mifflintown
dmh	Mahantango	sc	Clinton
doo	Onondaga–Old Port	skt	Keyser–Tonoloway
dtr	Trimmers Rock	swc	Wills Creek

(8) Select the desired aquifer file name (e.g., dc.dat).

(9) Click on OK. The Line Plot dialog box will reappear.

(10) Click on OK. A plot of the new aquifer data will be displayed on the graph.

Add title of new graph.

(1) Select the original title of the graph.

(2) Choose Edit | Delete.

(3) Follow the instructions for *Adding a Graph Title* which were described previously in this section.

Completing the Additional Graphs

Save and print the additional lognormal probability graphs by following the instructions presented previously.

Closing the Application and Quitting Windows

Before closing the application, be sure to save all active files.

(1) Choose File | Exit. If you have saved all active files, the application window closes. If you changed an active file but did not save it, then the Exit dialog box will appear.

(2) If the Exit dialog box appears, choose Yes to save the file. The Windows desktop will appear on the screen.

(3) Open another application or quit Windows and turn off the computer. Be sure to close all open applications before quitting Windows.

CHAPTER 14

Method for Preparing Groundwater Availability Maps

14.1 INTRODUCTION TO METHODS

A groundwater availability map is a type of hydrogeologic map which provides an estimate of the quantity of water available from the aquifers of a region. Groundwater availability may be commonly represented by median well yield, typical ranges of well yield, or other such measures. In this chapter, the groundwater availability of a given aquifer is expressed by a measure known as the *yield potential index.* This index is based on the frequency distribution of reported well yields of a chosen aquifer (see Chapters 12 and 13).

The purpose of this chapter is to prepare a map on which the relative availability of groundwater from each aquifer of a project region is illustrated by a unique map color. For example, an aquifer mapped as green exhibits high groundwater availability and is suitable for most uses from domestic to public supply to industrial. In contrast, an aquifer mapped as red is one that exhibits low groundwater availability.

This chapter describes a method for estimating potential well yield from lognormal graphs of frequency distributions (Chapter 13). The estimates of potential well yield are employed to prepare groundwater availability maps.

For users who have access to GRAPHER for Windows software, lognormal probability graphs of the well yields of the aquifers of the case study described in Chapter 10 may be created and printed by following the instructions given in Chapter 13.

167

14.2 ESTIMATING POTENTIAL WELL YIELD

Purpose

To estimate potential well yield of aquifers from frequency distribution graphs.

Equipment and Materials

- frequency distribution graphs of well yield (see Chapter 13)
- electronic calculator
- potential well yield worksheet (Figure 14.1)
- pencil (3H)

Procedure

List the aquifers of the region. Copy the names of the aquifers of a region under study (downward in order of youngest to oldest) onto the first column of the potential well yield worksheet (Figure 14.1).

Determine percentile measures of well yield.

(1) Using a frequency distribution graph of well yield from one of the aquifers listed, read the well yields that correspond to the 15, 50, and 85 percentiles of the graph and record the values in the appropriate columns next to the name of that aquifer.

(2) Repeat this step for all of the aquifers of the study region.

Calculate a yield potential index for each aquifer.

The yield potential index (Y_p) for each aquifer is calculated using the root–mean square equation, that is

$$Y_p = \sqrt{(Y_{15}^2 + Y_{50}^2 + Y_{85}^2)/3} \qquad (14.1)$$

where Y_p is the yield potential index, Y_{15} is the well yield at 15th percentile, Y_{50} is the well yield at 50th percentile, and Y_{85} is the well yield at 85th percentile.

The yield potential index is easily calculated on an electronic calculator. Follow the steps below.

(1) Enter the value of Y_{15}.
(2) Press [x²] ("square").
(3) Press [+].

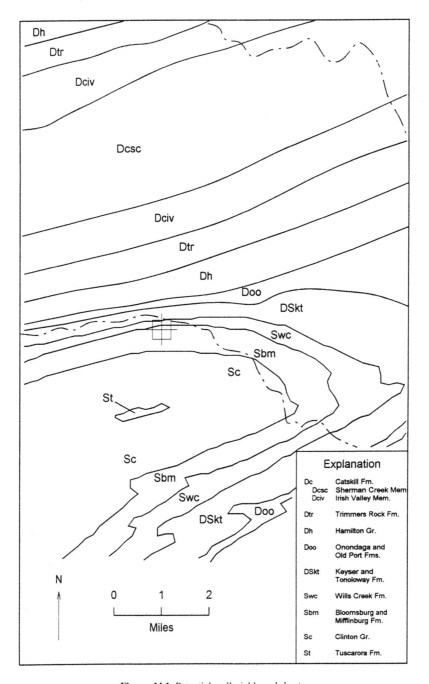

Figure 14.1 Potential well yield worksheet.

Explanation

Dc	Catskill Fm.
Dcsc	Sherman Creek Mem
Dciv	Irish Valley Mem.
Dtr	Trimmers Rock Fm.
Dh	Hamilton Gr.
Doo	Onondaga and Old Port Fms.
DSkt	Keyser and Tonoloway Fm.
Swc	Wills Creek Fm.
Sbm	Bloomsburg and Mifflinburg Fm.
Sc	Clinton Gr.
St	Tuscarora Fm.

N

0   1   2

Miles

(4) Enter the value of Y_{50}.

(5) Press [x^2].

(6) Press [$+$].

(7) Enter the value of Y_{85}.

(8) Press [x^2].

(9) Press [$=$].

(10) Press [$\div$].

(11) Enter 3.

(12) Press [$=$].

(13) Press [$\sqrt{}$] ("square root").

Record the resulting value in the Y_p (Yield Potential Index) column in Figure 14.1. Repeat the procedure until a yield potential index has been calculated for all of the aquifers and the values recorded on Figure 14.1.

14.3 PREPARING A GROUNDWATER AVAILABILITY MAP

Purpose

To construct a groundwater availability map.

Equipment and Materials

- potential well yield worksheet (e.g, Figure 14.1)
- groundwater availability base map (e.g., Figure 14.2)
- colored pencils
- pencil (3H)

Procedure

The procedures in this section assume that yield potential indexes have been calculated for all aquifers of a project region (see Section 14.2). Table 14.1 displays the map colors to be used, the ranges of yield potential index associated with each map color, and the suitability of a particular aquifer for several common uses.

Determine a map color for each aquifer. Using the information presented in Table 14.1, select the category of yield potential index and the associated map color for each of the formations of the project region.

Complete the groundwater availability map. On a base map of the

AQUIFER	WELL YIELD gpm			Y_p
	Y_{15}	Y_{50}	Y_{85}	

Figure 14.2 Groundwater availability base map (based on Berg and Dodge, 1981, p. 229).

TABLE 14.1. Yield Potential, Map Color, and Groundwater Availability.

Yield Potential Index (Y_p)	Map Color	Groundwater Availability		
		Domestic Use	Limited Public Supply	Public and Industrial Use
>50	Green	Excellent	Excellent	Good
30–50	Orange	Excellent	Good	Fair
10–29	Yellow	Good	Fair	Poor
<10	Red	Fair	Poor	Inadequate

project region (e.g., Figure 14.2), shade each of the aquifers in the map color that corresponds to its yield potential category.

14.4 REFERENCE

Berg, T. M. and Dodge, C. M., 1981. *Atlas of Preliminary Geologic Quadrangle Maps of Pennsylvania.* Pennsylvania Geological Survey Map 61.

CHAPTER 15

Method for Preparing
Test Well Site Maps

15.1 INTRODUCTION TO METHODS

As the discussion in Chapter 9 indicated, test wells should be drilled at sites which maximize the hydrogeologic factors that influence well yield. Thus, the identification of test well sites begins with assembling all of the information available on the hydrogeologic setting and groundwater availability of a region. The source materials particularly useful include topographic maps, geologic maps and columns, structure sections, aerial photographs and fracture trace maps, well yield frequency distributions, and groundwater availability maps.

After all test well sites have been identified, ranked, and plotted on a test well sites map, each (or the "best" ones) should be examined in detail by means of a geologic site survey and, if possible, a surface geophysical site survey. The results of these field surveys are employed to confirm or revise the priority ranking of test well sites. The drilling of the test wells begins at the site ranked #1, and proceeds down the priority list until the desired quantity of water is obtained.

15.2 CONSTRUCTION OF A TEST WELL SITES MAP

Purpose

To construct a test well sites map.

Equipment and Materials

- fracture trace map (e.g., Figure 11.4)
- groundwater availability map (e.g., Figure 14.2)
- test well site base map (e.g., Figure 15.1)
- test well site worksheet (e.g., Figure 15.2)
- light table
- pencil (3H)

Procedure

Select all potential test well sites situated on fracture traces.

(1) If available, place a fracture trace map (e.g., Figure 11.4) of the project region on a light table and overlay a test well site base map (e.g., Figure 15.1).

(2) Align the two maps so that all borders and features of the top map lie directly atop those of the bottom map, with the fracture traces clearly visible.

(3) On the test well site base map, mark the centers of each fracture trace lightly in pencil with a small circle.

(4) Label each of these sites with an identification code (e.g., MWA-1, MWA-2, etc.).

(5) Copy these well site ID codes to the test well site worksheet (Figure 15.2).

Determine groundwater availability of the aquifer at the site.

(1) Place a groundwater availability map (e.g., Figure 14.2) of the project region on a light table and overlay the test well sites base map (e.g., Figure 15.1).

(2) Align the maps so that all borders and features of the top map lie directly atop those of the bottom map.

(3) For each of the aquifers, select from Table 14.1 the descriptive term of groundwater availability (excellent, good, etc.) that applies to the type of use intended for the final production well (domestic, public, etc.).

(4) Record this term, next to the appropriate well code, in the second column of the test well site worksheet (Figure 15.2).

Summarize the hydrogeologic factors that influence well yield. For each of the well sites, record the major rock type, structural setting, approximate thickness of aquifer, and topographic setting in the appropriate columns.

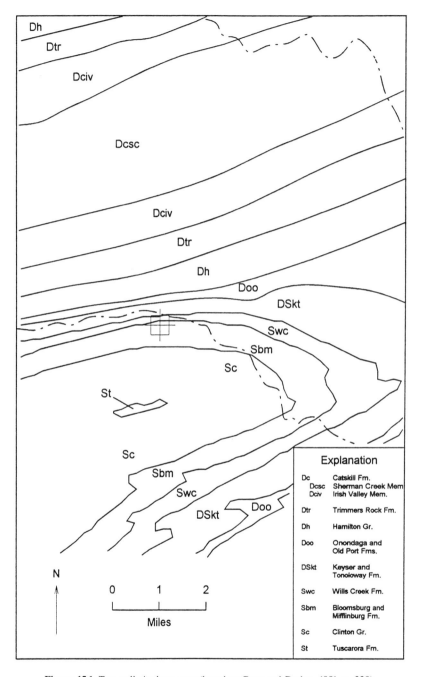

Figure 15.1 Test well site base map (based on Berg and Dodge, 1981, p. 229).

175

Well Site	Groundwater Availability	Rock Type	Structural Setting	Thickness of Aquifer	Topographic Setting	Rank

Figure 15.2 Test well sites worksheet.

Establish a site priority ranking.

(1) Evaluate all of the influencing factors associated with each of the well sites (see Figure 15.2) and, based on their relative yield potential, rank each site by assigning a priority number to it (e.g., 1 = best, 2 = next best, etc.).

(2) Enter the ranking number in the last column of the worksheet (Figure 15.2).

Mark and label the ranked sites.

(1) Draft a well symbol (e.g., ☼) on the base map of the project region (e.g., Figure 15.1) at each of the sites where test wells should be drilled.

(2) Assign each site a number that indicates its priority ranking (e.g., 1 = first choice, 2 = second choice, etc.).

15.3 REFERENCE

Berg, T. M., and Dodge, C. M., 1981. *Atlas of Preliminary Geologic Quadrangle Maps of Pennsylvania.* Pennsylvania Geological Survey Map 61.

PART 3

AQUIFER AND WELL HYDRAULICS

CHAPTER 16

Principles of Aquifer and Well Hydraulics

16.1 ELEMENTS OF PUMPING WELLS

G ROUNDWATER flow associated with pumping wells differs from that of one-directional flow in several essential characteristics. The important elements of pumping wells are defined below and are displayed graphically in Figure 16.1.

Discharge rate is the volume of water per unit time discharged from a well. It is measured commonly as pumping rate in gallons per minute (gpm). However, in order to maintain consistent units with other factors, discharge rate values should be expressed in units of L^3/T (e.g., 47 ft.^{3}/min.).

Static water level is the level at which water stands in a well when no water is being removed from the aquifer by pumping. It is expressed as the depth of the water level in the well below the ground surface (Driscoll, 1986, p. 206). In an unconfined aquifer, it represents the level of the water table at the well site, whereas in tightly cased wells drilled in confined aquifers, static water level represents the level of the potentiometric surface.

Pumping water level is the level at which water stands in a well when pumping is in progress (Driscoll, 1986, p. 206).

Drawdown is the difference between the water table or potentiometric surface and the pumping water level (Driscoll, 1986, p. 206). Its measurement units are those of length [L] (e.g., 32.6 ft.).

The *cone of depression* is the conical surface of the water level created in an unconfined aquifer by pumping water from a well (Environmental Protection Agency, 1976, p. 44). In many kinds of hydrogeologic investigations, and particularly in groundwater contamination problems, evaluation

181

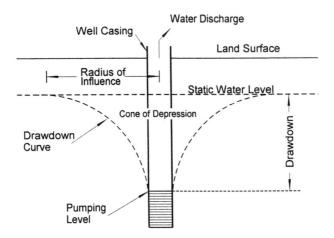

Figure 16.1 Elements of a pumping well (modified from Miller, 1980, p. 49).

of cones of depression is important because they represent an increase in hydraulic gradient, which in turn influences both the direction and velocity of groundwater flow. Because of differences in storage, the cones of depression in confined aquifers are larger than those in unconfined aquifers.

The *drawdown curve* is the profile line formed by the intersection of the surface of the water level around a pumping well and a vertical plane passing through the well. In an unconfined aquifer, the drawdown curve depicts the level to which the formation remains saturated. In a confined aquifer, it represents the hydrostatic pressure in the aquifer. At any given point on the drawdown curve, the difference between the water level indicated by the curve and the static water level is the drawdown in the aquifer.

The *radius of influence* is the horizontal distance from the center of a well to the limit of the cone of depression (Driscoll, 1986, p. 209). This limit is the point where there is no lowering of the water table (or potentiometric surface). Its measurement units are those of length [L] (e.g., 800 ft.).

16.2 AQUIFER PROPERTIES

Knowledge of the hydraulic properties of aquifers is essential. They provide an estimate of the water yield and permit the prediction of the amount of drawdown that will result from any given pumping rate and the shape and areal extent of the cone of depression. Two important hydraulic properties are *transmissivity* and *storativity*.

In Chapter 1 the volume of water that can be transmitted by a porous material was defined as the hydraulic conductivity. This term, however, describes the flow of water only through a unit cross-sectional area (e.g., 1 ft. × 1 ft.) of an aquifer, hence it fails to describe adequately the flow characteristics of an entire aquifer. The flow of water through an aquifer is described by transmissivity.

Transmissivity is defined as "the capacity of an aquifer to transmit water of the prevailing kinematic viscosity" (Heath, 1987, p. 26). It represents the rate at which water flows through a 1-foot wide vertical strip of aquifer, extending through the full saturated thickness, under a hydraulic gradient of 1 (i.e., 100%) (see Figure 16.2). This rate is expressed as $T = Kb$, where T is transmissivity, K is hydraulic conductivity, and b is aquifer

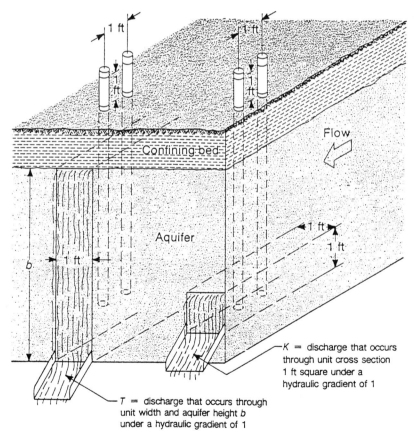

Figure 16.2 Illustration of hydraulic conductivity and transmissivity. (Reprinted from Driscoll, 1986, p. 210 with permission from Wheelabrator/Johnson Screens.)

thickness. Transmissivity may be considered the hydraulic conductivity of a unit width of the full thickness of the aquifer, and may be calculated by multiplying hydraulic conductivity by aquifer thickness. Hence, it has the dimensions of area per time $[L^2/T]$ (e.g., 14 ft.^{2}/min.). Aquifers that have transmissivity values greater than 10 ft.^{2}/min. show good potential for water well exploitation.

Storativity indicates how much water can be removed from an aquifer by pumping. Heath (1987, p. 28) defines it as "the volume of water that an aquifer releases from or takes into storage per unit change in head." For example, in a vertical column with a horizontal cross section of 1 square foot extending through the aquifer, storativity equals the volume of water released from or gained by the aquifer when the water table or potentiometric surface falls or rises 1 foot (see Figure 16.3). Storativity is expressed as the ratio $S = V'/V$, where V' equals the volume of water

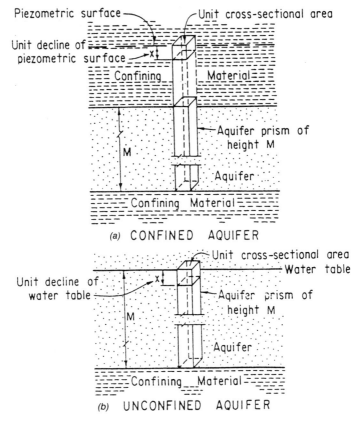

Figure 16.3 Illustration of storativity (from U.S. Dept. of the Interior, 1981, p. 26).

released and V is the volume of material drained in a water table aquifer, or the volume defined by the change in head for an artesian aquifer. Since $V'/V = L^3/L^3$, storativity is dimensionless. In unconfined aquifers storativity typically ranges between 0.1 and 0.3. The storativity of confined aquifers is substantially smaller and generally ranges between 0.0001 and 0.00001. [For additional information, see Driscoll (1986), pp. 76, 209–210, and U.S. Dept. of the Interior (1981), pp. 19–27.]

16.3 AQUIFER TESTS

An aquifer test is employed to determine the hydraulic properties of an aquifer (or other water-bearing material). It consists of the withdrawal of water from a well and the subsequent observation of the response of the aquifer or the well to this withdrawal.

During a typical aquifer test a well is pumped at a known and constant rate for a specified period of time and periodic measurements of the water level in the pumping well and in one or more nearby observation wells are taken. After the pump has been shut off at the conclusion of the test, measurements of the recovery of the water level are also taken. Figure 16.4 illustrates an example of an arithmetic plot of drawdown and recovery with respect to time in an observation well. The left portion of the plot depicts drawdown and the right portion depicts recovery. The dashed line represents the hypothetical extension of drawdown had pumping not been terminated.

Analysis of the results of aquifer tests provides important knowledge of

- aquifer properties
- drawdown in a well at future times and different discharge rates
- radius of the zone of influence for individual or multiple wells
- effect of new water withdrawals on existing wells
- nature and position of aquifer boundaries.

See Chapter 18 for additional information about aquifer tests.

16.4 FLOW TOWARD A PUMPING WELL

When a well is pumped, discharge water is first derived from casing storage and then from aquifer storage within the immediate vicinity of the well, and the water level in the well falls below the position of the static water level (i.e., the potentiometric surface or water table) outside the well. Because the water level is lower in the pumping well than at any place in the surrounding aquifer, a region of low pressure develops in the

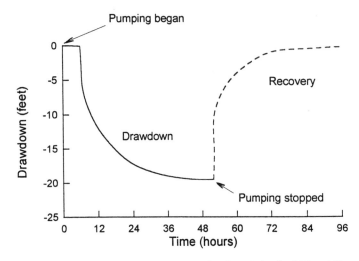

Figure 16.4 Drawdown-recovery plot (modified from Driscoll, 1986, p. 252).

aquifer near the well. Consequently, the water in the aquifer responds by flowing to the lower level in the well to replace that being withdrawn by the pump.

In an ideal confined aquifer, the flow is radial; that is, groundwater moves toward the well from every direction, as if moving along the spokes of a wheel toward the hub (see Figure 16.5). If the well fully penetrates the aquifer and the potentiometric surface is not drawn below the bottom of the confining bed, the flow lines to the well are horizontal and parallel (see Figure 16.6).

The pressure — that is, the force — that drives the water toward the well is the difference in *hydraulic head,* which is the difference between the water level inside the well and the water level at any place in the aquifer outside the well $(h_1 - h_2)$. An increase in head produces a corresponding increase in hydraulic gradient. Moreover, as the hydraulic gradient increases, velocity increases as the flow converges toward a well, because the area through which the flow is occurring gets progressively smaller. This phenomenon is in accordance with Darcy's law,

$$Q = KA(h_1 - h_2)/L \qquad (16.1)$$

which states that the rate of flow varies directly with the hydraulic gradient. As a result, the lowered potentiometric surface develops a continually steeper slope toward the well and takes the form of an inverted cone centered at the well (i.e., the cone of depression).

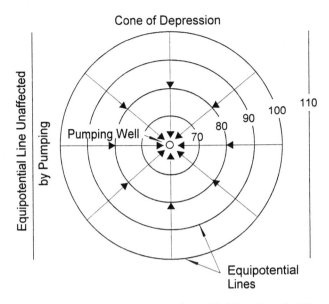

Figure 16.5 Radial flow toward a pumping well (modified from Caswell, 1979, p. 29).

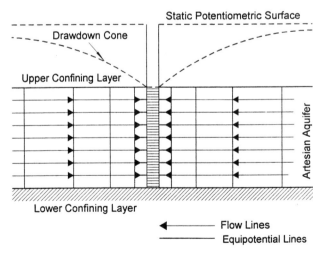

Figure 16.6 Flow to a pumping well in an ideal confined aquifer (modified from U.S. Dept. of the Interior, 1981, p. 47).

In an unconfined aquifer, flow lines to the well do not remain horizontal and parallel because water is derived from gravity drainage of the aquifer materials around the well. Instead, there are both horizontal and vertical flow components.

Other factors, as well, can cause the flow to pumping wells to deviate from the strictly radial flow of ideal confined aquifers. For example, flow toward wells that only partially penetrate the aquifer includes a vertical flow component. As a result, the amount of drawdown in the vicinity of a partially penetrating well is greater than that around a fully penetrating one. In this case, increased drawdown is associated with decreased percentage of open well. Additionally, many aquifers are bounded by leaky confining beds that release water into the aquifer in response to pumping. Such an aquifer exhibits the flow characteristics of both a confined and an unconfined aquifer. That is, it exhibits both horizontal and vertical components of flow. [For additional information, see Driscoll (1986), pp. 249–250, and U.S. of the Dept. Interior (1981), pp. 45–49.]

16.5 DRAWDOWN IN IDEAL AQUIFERS

All pumped wells are surrounded by a cone of depression, the size and shape of which depends on such factors as the initial slope of the water table, pumping rate, pumping duration, and aquifer properties. In two dimensions, the drawdown in any radial direction from the pumping well forms a smooth curve (the drawdown curve, see Figure 16.7). At a point

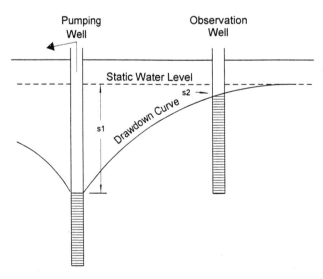

Figure 16.7 Drawdown curve (modified from Caswell, 1979, p. 36).

some distance from the well the water level is essentially unaffected by the pumping. For a given well and aquifer this distance depends on both the pumping rate and the length of time the well is pumped.

As pumping continues, the cone of depression expands as more water must be derived from aquifer storage at increasingly greater distances from the well. The radius of influence increases as the cone expands, and drawdown at any point also increases as the cone deepens to provide the additional head required to move the water from greater distances. The rate of expansion is not constant, however. Instead, the cone expands and deepens more slowly with time, because an increasing volume of stored water is available with each additional foot of horizontal expansion. The cone of depression will continue to expand until it intercepts enough of the flow in the aquifer to equal the pumping rate. When the cone has stopped expanding, equilibrium exists. In some wells, equilibrium occurs within a few hours after pumping begins; in others, however, it may take years, if it is ever attained.

From the discussion above it should be evident that the drawing down of the water table of an unconfined aquifer or the potentiometric surface of a confined aquifer is one of the most obvious and most important hydrogeologic effects of a pumping water well. The amount of drawdown around a pumping well has particular significance for the successful development of water supply wells and is related to several factors. These include

- discharge rate
- time since pumping began
- distance from the pumping well
- transmissivity
- storativity

The clear and definite relationship between drawdown and these factors provides the basis for predicting the amount of drawdown resulting from the withdrawal of water from an aquifer.

The relationships between the amount of drawdown and the various hydraulic factors can be summarized by the following rule:

> Drawdown in the vicinity of a pumping well is directly proportional to the pumping rate and the time since pumping began and is inversely proportional to the transmissivity, the storativity and the square of the distance between the pumping well and any point on the cone of depression.

These relationships can be expressed mathematically in a set of well discharge equations, which provide a means for determining the hydraulic properties of an aquifer (i.e., hydraulic conductivity, transmissivity, and storativity).

There are two basic equilibrium equations: one for unconfined conditions and the other for confined conditions (for details see Driscoll, 1986, p. 214). The rate of discharge from a pumping well that has reached equilibrium was described by Thiem (1906). For equilibrium (or steady-state conditions), the *Thiem equation* is appropriate and allows for the determination of the hydraulic conductivity of an aquifer. For nonequilibrium (or nonsteady-state) conditions, the *Theis equation* may be employed to determine the transmissivity and storativity of an aquifer. For both equations, however, all conditions in the well and aquifer are assumed to be in equilibrium; that is, well discharge is constant, drawdown and radius of influence have stabilized, and water enters the well in equal volumes from all directions. Additionally, both equations assume that horizontal flow occurs everywhere in the aquifer, with recharge occurring at the edge of the cone of depression. In practice, these assumptions are severe and limit the utility of the Thiem equations.

Theis (1935) developed an equation that permits the determination of the hydraulic properties of an aquifer before equilibrium conditions are reached. The assumptions on which the equation is based, however, are strict and seldom met in practice. These assumptions (from Todd, 1980, p. 124) are

(1) The aquifer is homogeneous, isotropic, of uniform thickness, and of infinite aeral extent.
(2) Before pumping, the potentiometric surface is horizontal.
(3) The well is pumped at a constant discharge rate.
(4) The pumped well penetrates the entire aquifer, and flow is everywhere horizontal within the aquifer to the well.
(5) The well diameter is infinitesimal so that storage within the well can be neglected.
(6) Water removed from storage is discharged instantaneously with decline of head.

Nevertheless, the equation has found wide application and is useful if the results are interpreted with due caution.

The Theis equation relates the amount of drawdown at a given distance from a pumping well to pumping rate, transmissivity, and storativity. In its simplest form the Theis equation is

$$s = \frac{QW(u)}{4\pi T} \qquad (16.2)$$

where s is the drawdown at any point in the vicinity of a pumping well [L],

Q is the rate of discharge [L³/T], T is the transmissivity [L²/T], $W(u)$ is known as the "well function of u," and

$$u = \frac{r^2 S}{4Tt} \tag{16.3}$$

where r is the distance from the pumping well to a point where drawdown is measured [L], S is the storativity [dimensionless], and t is the time since pumping began [T].

Aquifers that satisfy all of the assumptions inherent in the Theis equation are known as *ideal aquifers*. Nonideal aquifers, such as leaky or unconfined aquifers, do not satisfy some of the important assumptions inherent in the Theis nonequilibrium equation. Analysis of the hydraulic properties of these kinds of aquifers requires modification of the equation (see Chapters 23 and 24).

16.5.1 DISCHARGE–DRAWDOWN RELATIONSHIP

With all other factors constant, drawdown at any point in the cone of depression is directly proportional to well discharge. In theory, this means that if well discharge is doubled, the amount of drawdown is doubled (see Figure 16.8).

16.5.2 TIME–DRAWDOWN RELATIONSHIP

For nonequilibrium conditions, drawdown varies directly with the logarithm of time since pumping began. Thus, if 10 minutes of pumping produces 15 feet of drawdown in a pumping well, then 100 minutes of pumping will result in 20 feet of drawdown, and 1,000 minutes will produce 25 feet of drawdown (see Figure 16.9).

16.5.3 DISTANCE–DRAWDOWN RELATIONSHIP

Previous discussion of the flow of groundwater toward a pumping well demonstrated that the slope of the sides of the cone of depression is steep near the well and becomes increasingly more gentle away from the well (see Figure 16.10). At a given discharge rate, the drawdown at any point on the cone of depression is inversely proportional to the log of the distance from the pumping well (see Figure 16.10). For example, if drawdown at a distance of one foot from the pumping well is 23 feet, then (depending on the transmissivity and storativity of the aquifer) the amount of drawdown at a distance of 10 feet might be 15 feet, and at a distance of 100 feet, 7 feet.

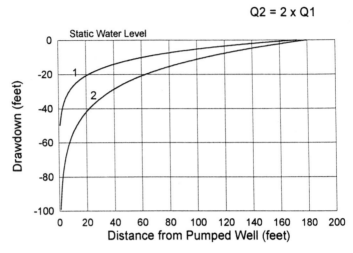

Figure 16.8 Discharge-drawdown relationship (based on U.S. Dept. of the Interior, 1981, p. 51).

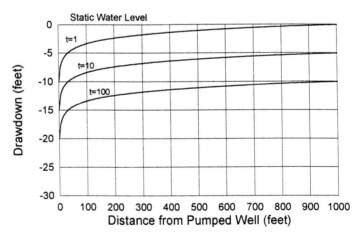

Figure 16.9 Time-drawdown relationship (based on U.S. Dept. of the Interior, 1981, p. 51).

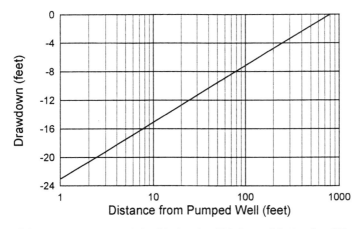

Figure 16.10 Distance-drawdown relationship (based on U.S. Dept. of the Interior, 1981, p. 90).

16.5.4 TRANSMISSIVITY–DRAWDOWN RELATIONSHIP

Figure 16.11 illustrates the influence of transmissivity on drawdown. At a given discharge, drawdown decreases with increased values of transmissivity, and the cone of depression takes a flatter shape with increased transmissivity. In an aquifer with low transmissivity, the cone is deep with steep sides and has a small radius. In an aquifer with high transmissivity, the cone is shallow with flat sides and has a large radius.

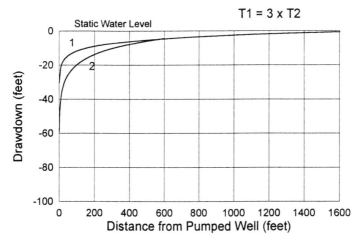

Figure 16.11 Influence of transmissivity on drawdown (based on U.S. Dept. of the Interior, 1981, p. 52).

The explanation for these different cone shapes is clear, for greater hydraulic head is required to move water through a less permeable formation than through a more permeable formation.

16.5.5 STORATIVITY–DRAWDOWN RELATIONSHIP

Drawdown is inversely proportional to storativity; thus, increased values of storativity result in decreased drawdown (see Figure 16.12). Like transmissivity, storativity also affects the shape of the cone of depression. For a confined aquifer storativity is small, and the pumping affects a relatively large area. [For additional information, see Driscoll (1986), pp. 207–212, and U.S. Dept. of the Interior (1981), pp. 43–53.]

16.6 EFFECTS OF LATERAL BOUNDARIES

One of the assumptions underlying the Theis equation is that the aquifer being analyzed is infinite in extent. As Heath (1987, p. 46) reminds us, "Obviously, no such aquifer exists on Earth." All aquifers are bounded, both *stratigraphically* and *laterally*. Aquifers are bounded stratigraphically by confining layers. Lateral boundaries affect the response of an aquifer to water withdrawals and fall into two types: *recharge* and *impermeable*.

A recharge boundary is a boundary along which flow lines originate and which typically serves as a source of recharge to the aquifer. A significant feature of a recharge boundary is that water withdrawals from the aquifer

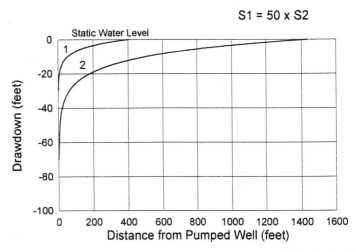

Figure 16.12 Influence of storativity on drawdown (based on U.S. Dept. of the Interior, 1981, p. 52.

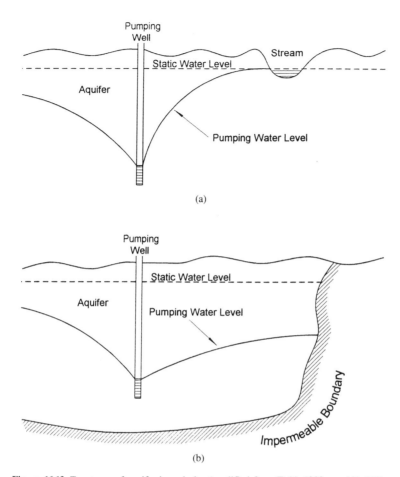

Figure 16.13 Two types of aquifer boundaries (modified from Todd, 1980, pp. 141, 145).

do not produce drawdown across the boundary. A typical example of a recharge boundary is a perennial stream [see Figure 16.13(a)]. In the case of a perennial stream situated within the radius of influence of a pumping well, water from the stream will induce recharge of the aquifer.

Therefore, when an aquifer test is conducted near a recharge boundary, drawdowns in an observation well will be less than that of an "infinite" aquifer, and the resultant time–drawdown curve will appear to be flatter. Initially, the rate of drawdown is influenced only by the water in the aquifer in the vicinity of the pumping well; but when the expanding cone of depression around the well intersects the recharge boundary, the additional supply of induced stream water results in reduced drawdown.

An impermeable boundary is a boundary that flow lines do not cross and exists where aquifers terminate against impermeable rock material. Examples of impermeable boundaries include the interface of a sand aquifer and a laterally adjacent clay bed and the bedrock wall of an alluvial valley [see Figure 16.13(b)].

When an aquifer test is conducted near an impermeable boundary, drawdowns in an observation well will be greater than that of an "infinite" aquifer and the time–drawdown curve will appear to be steeper. Initially, the time–drawdown curve resembles that of an "infinite" aquifer, but when the expanding cone of depression around the pumping well intersects the impermeable boundary, no additional water can be derived from that direction. Consequently, the cone must expand and deepen in the other directions, and the effect on the time–drawdown curve is a steepened slope.

16.7 LEAKY CONFINING BEDS

The assumption inherent in the Theis equation that all of the water discharged from the pumping well is derived solely from storage in the aquifer is rarely satisfied under actual field conditions. The reason for this is that geologic formations overlying and underlying a confined aquifer are never completely impermeable. The common case is that an aquifer is only one member of a multiple-aquifer system in which a succession of aquifers are separated by intervening low-permeability confining layers. Typical examples of these systems occur in alluvial valleys, plains, or former lake basins where permeable sands and gravels are interbedded with impermeable clays.

Even when a well is only screened in a single aquifer, this aquifer commonly receives appreciable inflow from adjacent strata. If the direction of the hydraulic gradient is favorable, semipermeable beds, either above or below the aquifer, can leak water into the aquifer. Hence, such an aquifer is called a leaky aquifer, although in actuality it is the confining bed that is leaky (see Figure 16.14).

A leaky confining bed may be considered a type of boundary. Aquifers that are bounded by leaky confining beds, which release water into the aquifer in response to withdrawals, produce drawdown patterns different from those that would be predicted by the Theis equation. When a so-called leaky aquifer is pumped, water is withdrawn both from the aquifer and from the saturated portion of the overlying confining layer. Lowering the potentiometric head in the aquifer by pumping produces a hydraulic gradient with the confining bed; consequently, groundwater moves vertically downward into the aquifer. The quantity of water moving downward

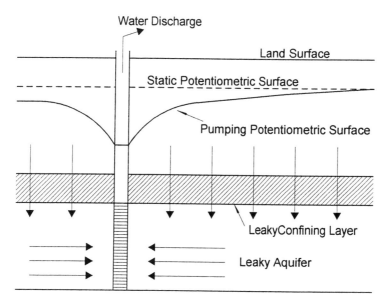

Figure 16.14 Illustration of a leaky aquifer system (modified from Todd, 1980, p. 138).

is proportional to the difference between the water table and the potentiometric head. As in the case of lateral boundaries, when the area of influence has expanded sufficiently so that the amount of increased seepage into aquifer equals the pumping rate, the discharge–drawdown relationship stabilizes.

Similarly, when a water table aquifer overlying a confined aquifer is pumped, the decline of the water table in the cone of depression reduces pressure and increases upflow from the confined aquifer.

16.8 EFFECTS OF ANISOTROPY

Anisotropy refers to an aquifer characteristic in which flow conditions vary with direction. (Anisotropic aquifers were introduced in Chapter 1.) In granular materials such as gravels, textural differences resulting from compaction may cause hydraulic conductivity to be less in the vertical direction than in the horizontal direction. In bedrock, anisotropy is almost always the result of the size, shape, orientation, and spacing of fractures. Where such anisotropy exists, the resultant effect is a distortion of drawdowns in the vicinity of a pumping well.

Commonly, pumping a well that has been sited on a fracture trace creates a zone of influence elongated in the direction of the long axis of

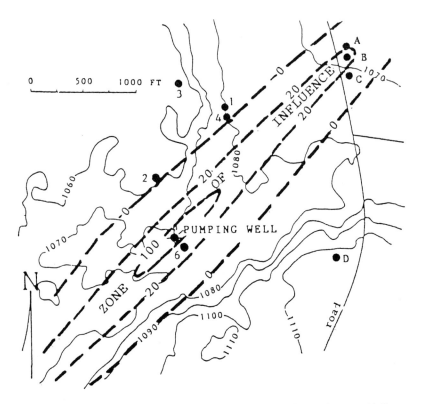

Figure 16.15 Effects of anisotropy produced by fracture concentration on the zone of influence of a pumping well. (Reprinted from Caswell, p. 1987, p. 33; copyright © 1987. *Water Well Journal.*)

that trace (see Figure 16.15). This effect results from hydraulic conductivities that are higher in the direction of the fracture trace than across it.

16.9 REFERENCES

Caswell, B., 1987. Anisotropy in fractured bedrock aquifers: It can be helpful. *Water Well Jour.*, Sept., p. 33.

Driscoll, F. G., 1986. *Groundwater and Wells* (2nd ed.). St. Paul, MN: Johnson Division.

Environmental Protection Agency, 1976. *Manual of Water Well Construction Practices.* Washington, D.C.: U.S. Govt. Printing Office.

Heath, R. C., 1987. *Basic Ground-Water Hydrology.* U.S. Geological Survey Water-Supply Paper 2220.

Miller, D. W. (ed.), 1980. *Waste Disposal Effects on Ground Water.* Berkeley, CA: Premier Press.

Theis, C. V., 1935. The relation between the lowering of the piezometric surface and the rate and duration of discharge of a well using ground water storage. *Trans. Amer. Geophysical Union.*, pp. 518–525.

Thiem, G. 1906. *Hydrologische Methoden.* Leipzig: J. M. Gephardt.

Todd, D. K., 1980. *Groundwater Hydrology.* New York, NY: John Wiley & Sons.

U.S. Dept. of the Interior, 1981. *Ground Water Manual.* Washington, D.C.: U.S. Govt. Printing Office.

CHAPTER 17

Case Study: Analysis of the Hydraulic Properties of an Aquifer

17.1 STATEMENT OF THE PROBLEM

A village water authority needs to increase its water supply by 500,000 gallons per day. It has authorized the drilling of a test well, for the purpose of determining if the required quantity of water can be obtained safely and economically. In order to proceed with the development of the new water supply system, the ability of the aquifer to transmit and store water must be analyzed.

17.2 PURPOSE AND SCOPE

The purpose of the project is to analyze the ability of the aquifer to yield the required quantity of water safely. The specific objectives are to

- conduct a pumping (i.e., aquifer) test on an exploratory well
- estimate the transmissivity and storativity of the aquifer

17.3 BACKGROUND AND PROJECT SETTING

An exploratory test well (designated Well No. 1) that was identified during the groundwater exploration phase of the project has been drilled at a site near the village (see Figure 17.1). In order to measure the effects of drawdown created by the pumping well during the aquifer test, an observation well (designated Well No. 2) has been drilled 400 feet away. Both wells are fully penetrating and are screened the full thickness of the aquifer. Additional specifications for these wells are given in Table 17.1.

201

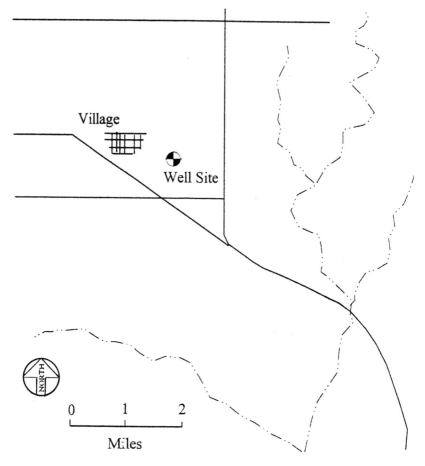

Figure 17.1 Test well site map.

TABLE 17.1. Well Specifications for Pumping Test of the Case Study.

Well	Diameter (in.)	Depth (ft.)	Screened Interval (ft.)	Distance from Well No. 1 (ft.)
No. 1	12	350	260–287	NA
No. 2	6	320	260–287	400

The village is situated in the midwestern United States within the till plains of the Central Lowlands physiographic region. In the immediate vicinity of the well sites, approximately 260 feet of clay overlie a sand and gravel aquifer 27-feet thick. This aquifer lies stratigraphically atop a thick layer of shale bedrock of low conductivity. Water levels in the wells of the region typically occur at depths of 65–70 feet.

17.4 METHODS OF INVESTIGATION

In order to estimate the hydraulic properties of the aquifer, the investigative methods described below will be employed.

TABLE 17.2. An Example of an Outline for an Aquifer Test Report.

Section	Products and Exhibits
TITLE PAGE	
1.0 INTRODUCTION	
1.1 Purpose and Scope	
1.2 Project Setting	Index map of project site
2.0 METHODS OF INVESTIGATION	
2.1 Well Test	
2.2 Data Analysis and Presentation	Map of well site(s)
3.0 RESULTS OF INVESTIGATION	
3.1 Analysis of Drawdown	Table(s) of drawdown data
	Time–drawdown graph(s)
3.2 Hydraulic Properties of Aquifer	Theis type graph
	Cooper-Jacob graph
	(Hantush graph)
	(Neuman graph)
3.3 Predicted Drawdown Effects	Graph/map of cone of depression
4.0 CONCLUSIONS AND RECOMMENDATIONS	
5.0 REFERENCES	

(1) Pumping of the test well (No. 1) at a constant discharge rate of 200 gpm for a period of 24 hours and measurement of the depth to water level in an observation well with an electric sounder at specified times (see Chapter 18)

(2) Calculation from the field measurements of drawdown and elapsed time since pumping began (see Chapter 19)

(3) Construction of a time–drawdown graph from aquifer test data (see Chapter 20)

(4) Determination of aquifer properties by means of the following methods: Theis method (see Chapter 21) and Cooper-Jacob method (see Chapter 22). (The Hantush-Jacob method of Chapter 24 and the Neuman method of Chapter 25 are optional.)

17.5 REPORT OF FINDINGS

The results of the investigation of aquifer properties can be summarized in a report. This report may take the form seen in Table 17.2.

CHAPTER 18

Field Procedure for Conducting Aquifer Tests

18.1 INTRODUCTION TO METHODS

A N aquifer test provides an effective means for determining the hydraulic properties of aquifers (see Section 16.3 for a background discussion of this topic). It involves the withdrawal of water from a control well at a specified rate and the measurement of drawdown in one or more observation wells.

As Driscoll (1986, p. 531) points out, "Pumping tests will not produce accurate data unless the tests are carried out methodically, carefully recording the time, discharge, and depth measurements." The accuracy of data obtained from an aquifer test depends on the following:

- maintaining a constant yield during the test (or periods of the test)
- measuring the drawdown and recovery carefully
- taking drawdown and recovery readings at appropriate time intervals
- selecting, according to the type of aquifer, the appropriate length of time for the test
- guarding against errors that arise from well interference, inflow from streams, and changes in barometric pressure

The minimum measurements required for any kind of pumping test include

- static water level
- time pumping began
- pumping (or discharge) rate
- depth to water level

- clock time of each water level measurement
- distance from the pumping well to observation wells

Table 18.1 displays the recommended time intervals for measuring drawdown in wells. All measurements are recorded on an aquifer test data form similar to Figure 18.1.

Although aquifer tests are more effective if water levels are measured in observation wells (i.e., multiple well tests), it is possible to obtain useful data from a production well (or monitoring well) where observation wells are not available. This method is called a *single well test*. Despite some loss of information (storativity, for example, cannot be determined directly), the method offers obvious economic advantages and is commonly employed in hydraulic testing carried out at waste disposal sites. Single well pumping tests are usually of two kinds: *recovery tests* and *slug tests*.

As the name implies, a recovery test involves shutting off the pump and observing the recovery of water levels in the same well. Because turbulence and well loss in a pumping well make the accurate measurement of water levels difficult, a recovery test provides a useful alternative to a drawdown test. The only difference in aquifer response between a drawdown test and a recovery test is the direction of change in water level.

TABLE 18.1. Recommended Time Intervals for Measuring Drawdown during a Pumping Test.

Drawdown in Observation Wells	
Time since Pumping Started (or Stopped) (min)	Time Intervals between Measurements (min)
0–60	2
60–120	5
120–240	10
240–360	30
360–1440	60
1440–termination	480
Drawdown in Pumping Well	
Time since Pumping Started (or Stopped) (min)	Time Intervals between Measurements (min)
0–10	0.5–1.0
10–15	1
15–60	5
60–300	30
300–1440	60
1440–termination	480

Pumped well:					Observation well:			
Diameter:					Diameter:			
Depth:					Depth:			
Well location:					Distance to pumped well:			
Time Data			Water Level Data			Discharge Data		Remarks
Pump on: _____			Static water level: _____			Measurement		
Pump off: _____			Elevation of			method:		
			measuring point: _____					
1	2	3	4	5	6	7	8	Page 1 of
Date	Clock time	Time since pumping began t	Time since pumping stopped t'	Depth to water	Draw-down s	Recovery s'	Dis-charge rate Q	

Figure 18.1 Aquifer test data form.

A slug test is another common method employed in field practice. Its advantages are that it can be used in low-conductivity materials and requires no pumping equipment. These characteristics are especially valuable in site investigations of waste disposal facilities. The slug test involves either injecting into or withdrawing a slug of water of known volume from the well or borehole, with an original head above its static level. The rate at which the water level rises or falls in the well is used to estimate the hydraulic properties of the aquifer. Although the injection procedure is most common, the slug withdrawal procedure is necessary in the investigation of petroleum spills or leaks, where the well or borehole is screened above the water table.

The selection of a particular aquifer test method is based on conformance of the site hydrogeology to the assumptions of the analytical solution test method. For example, an anisotropic, homogeneous, extensively confined, nonleaky aquifer, which may be analyzed by the nonequilibrium method of Theis or Cooper-Jacob, permits the use of a constant discharge, multiple-well aquifer test method. In any case, knowledge of the hydrogeology of the test site is imperative. [For additional information, see Driscoll (1986), pp. 534–554, and Heath (1987), pp. 34–35, 42–43.]

18.2 CONSTANT RATE–MULTIPLE WELL METHOD

Purpose

To conduct a constant rate–multiple well aquifer test.

Reference

Driscoll, F. G., 1986. *Groundwater and Wells.* (2nd ed.) St. Paul, MN: Johnson Division, pp. 534–554.

U.S. Dept. of the Interior, 1981. *Ground Water Manual.* New York, NY: John Wiley & Sons, pp. 225–246.

Equipment and Materials

- pumping system
- discharge-rate measurement device, or container of known volume (e.g., 3 gal.)
- stopwatch
- water-level measurement equipment (see Chapter 3)
- aquifer test data forms (e.g., Figure 18.1)
- pencils

Procedure

(1) Install observation well(s) or identify existing wells which can be used to measure drawdown during the pumping test. Ideally, three observation wells should be employed. Well spacing should be logarithmic and designed to provide at least one logarithmic cycle of distance–drawdown data. A typical spacing is 100, 400, and 1,000 feet (Walton, 1987, p. 10).

(2) Determine the length of time of the pumping test. Typically, a constant-rate test should last 24 hours on a confined aquifer and 72 hours on an unconfined aquifer.

(3) Evaluate the potential interference of pumping the test well on other wells in the vicinity.

(4) Install pumping equipment (a well-driller subcontractor can perform this task).

(5) Arrange for proper disposal of discharge water far enough away from the well to avoid recharge. Contaminated water will require special handling.

(6) At least one day prior to the test, pump the control well for several hours in order to evaluate such factors as maximum drawdown, pump performance, proper discharge measurement method, and response of observation wells.

(7) Stop the pump and allow the water level in the well to recover to its static water level. Normally overnight will suffice, but 24–72 hours may be required in some cases.

(8) Assemble all discharge and water-level measurement equipment and data forms at the site.

(9) Record the description of well and project information on an aquifer test data form (e.g., Figure 18.1).

(10) Begin the test by measuring and recording the static water level in the pumping well and observation well(s). The electric sounder method for measuring depth to water level in wells is described in Chapter 3.

(11) Noting the time, begin pumping at the established discharge rate. Various discharge measurement methods are described by Driscoll (1986, pp. 536–547). A simple and reasonably accurate method for measuring relatively low discharge rates is to observe the time required to fill a container of known volume. The discharge rate is found by

$$Q = 60 \ (V/t) \qquad (18.1)$$

where Q is the discharge rate, in gpm; V is the volume of the container, in gal.; and t is the time to fill the container, in sec. Be careful to maintain a constant yield during the test.

(12) Measure the drawdown in the observation well(s) at the appropriate time intervals (see Table 18.1).

(13) At the end of the required test period, turn off the pump and record the time.

(14) Measure and record recovery levels, following the same time intervals as for drawdown levels.

(15) Before leaving the well site, make sure all necessary data have been recorded on the aquifer test data forms.

18.3 REFERENCES

Driscoll, F. G., 1986. *Groundwater and Wells.* (2nd ed.). St. Paul, MN: Johnson Division, pp. 534–554.

Heath, R. C. 1987. *Basic Ground-Water Hydrology.* U.S. Geological Survey Water-Supply Paper 2220.

U.S. Dept. of the Interior, 1981. *Ground Water Manual.* New York, NY: John Wiley & Sons, pp. 225–246.

Walton, W. C., 1987. *Groundwater Pumping Tests.* Chelsea, MI: Lewis Publ., Inc.

CHAPTER 19

Analytical Procedure for Determining Time and Drawdown from Pumping Test Measurements

19.1 INTRODUCTION TO METHODS

IN order to be useful for estimating aquifer properties, aquifer test measurements of depth to water in well and clock time must be transformed into data indicating drawdown and time since pumping began (time and drawdown relations were introduced in Chapter 16). If recovery is also measured during the test, then these measurements must likewise be transformed.

This section describes two methods by which the transformation of data may be carried out. They are the *Electronic Calculator Method* (Section 19.2) and the *Microcomputer and Electronic Spreadsheet Method* (Section 19.3).

Figure 19.1 displays a field plot of aquifer test data for observation well T9-1 of the case study in Chapter 17. This field plot includes both drawdown and recovery data. Use these measurements (or your own aquifer test measurements) for application of the methods described in this section.

19.2 ELECTRONIC CALCULATOR METHOD

Purpose

To calculate drawdown and time since pumping began from pumping test measurements.

Pumped well: Test No. 1					Observation well: Test No. 2			
Diameter: 12 inches					Diameter: 6 inches			
Depth: 350 feet					Depth: 320 feet			
Well location:					Distance to pumped well: 400 feet			
Time Data			Water Level Data			Discharge Data		Remarks
Pump on: 7/17/94 0800 Pump off: 7/18/94 0800			Static water level: 67.8 feet Elevation of measuring point: _____			Measurement method:		
1	2	3	4	5	6	7	8	Page 1 of 2
Date	Clock time	Time since pumping began t	Time since pumping stopped t'	Depth to water	Draw- down s	Recovery s'	Dis- charge rate Q	
7/16	0800			67.68				
	1200			67.70				
	1600			67.72				
	2000			67.74				
	2400			67.76				
7/17	0400			67.78				
	0800			67.80			200 gpm	pump on
	0803			69.6				
	0804			70.1				
	0805			70.4				
	0806			70.7				
	0810			71.7				
	0812			72.1				
	0814			72.4				
	0818			72.9				
	0824			73.5				
	0830			74.0				
	0840			74.5				
	0900			75.4				
	0920			76.0				
	0940			76.5				
	1000			77.0				
								cont. on p. 2

Figure 19.1 Aquifer test data (Test No. 2).

Pumped well: Test No. 1			Observation well: Test No. 2					
Diameter: 12 inches			Diameter: 6 inches					
Depth: 350 feet			Depth: 320 feet					
Well location:			Distance to pumped well: 400 feet					

Time Data			Water Level Data		Discharge Data			Remarks
Pump on: 7/17/94 0800			Static water level: 67.8 feet		Measurement			
Pump off: 7/18/94 0800			Elevation of		method:			
			measuring point: _____					

1	2	3	4	5	6	7	8	Page 2 of 2
Date	Clock time	Time since pumping began t	Time since pumping stopped t'	Depth to water	Draw-down s	Recovery s'	Dis-charge rate Q	
7/17	1030			77.5				
	1100			77.7				
	1200			78.4				
	1300			78.9				
	1400			79.2				
	1600			79.8				
	1800			80.3				
	2000			80.7				
	2400			81.3				
7/18	0400			81.8				
	0800			82.2				pump off

Figure 19.1 continued. Aquifer test data (Test No. 2).

Equipment and Materials

- aquifer test form and test data (e.g., Figure 19.1)
- electronic calculator
- pencil (3H)

Procedure

Note: For the pumping test from the case study in Chapter 17, use the data given in Figure 19.1. If you have your own project data, make a copy of the aquifer test form in Figure 19.1 and record your data on that copy.
Calculate the time since pumping began.

(1) For each water level measurement subtract the pump on time from the clock time (in column 2).
(2) Enter the values on the aquifer test data form in column 3.

Calculate the drawdown.

(1) Subtract the static water level from the depth to water (in column 5) of each drawdown measurement.
(2) Enter the drawdown values on the aquifer test data form in column 6.

Calculate the recovery (optional, if data are available).

(1) Subtract the static water level from the depth to water (in column 5) for each recovery measurement.
(2) Enter any recovery values (if available) in column 7.

19.3 MICROCOMPUTER AND ELECTRONIC SPREADSHEET METHOD

Purpose

To calculate drawdown and time since pumping began from aquifer test measurements.

Equipment and Materials

- aquifer test data (e.g., Figure 19.1)
- microcomputer with an 80386 processor or higher, MS-DOS 3.3 or higher, Windows 3.1 or higher, hard drive, floppy disk drive, and graphics printer

- Windows version of electronic spreadsheet software (e.g., Lotus 1-2-3, Quattro Pro, or Excel)
- Hydrodata Diskette

Procedure

This method employs electronic spreadsheet software to transform aquifer test measurements. Field measurements (use Figure 19.1 or your own project data) are entered into a spreadsheet template (filename: fplot.txt), which is loaded from the Hydrodata Diskette. The computations are accomplished by means of formulas entered into the spreadsheet template. When the calculations are completed, a table of the time–drawdown data may be printed.

Starting Up

Turn on the computer, monitor, and printer. Wait until the Windows desktop is displayed.

Open the application.

(1) Open the Windows group icon that contains your software application (e.g., Quattro Pro), or click on the Windows 95 Start button and point first to Programs and then to the application folder.

(2) Double click on the application icon from the group window, or click on the name of the application from the drop-down list.

Opening the Aquifer Test Water Data Template File

(1) Insert the Hydrodata Diskette into drive A (or B).

(2) Choose File | Open. The Open File dialog box will be displayed on the screen.

(3) Click on the arrow next to File Type list box and select text files (e.g., *.txt) from the drop-down list.

(4) Specify the drive that contains your file by clicking on the arrow next to the Drives list box and then clicking on the drive name (e.g., A).

(5) Specify the file you want to open by selecting the name of the desired file in the File Name drop-down box (e.g., fplot.txt).

(6) Initiate file opening.

For Quattro Pro and Lotus 1-2-3: Click on OK
For Excel: The Text Import Wizard will appear. Click twice on Next > and once on Finish.

The aquifer test data template will be displayed on the monitor screen.

Entering Spreadsheet Formulas

Note: For Microsoft Excel users, enter an equal sign (=) before each of the formulas appearing below.

Enter the formulas for calculating discharge rate (Q).

(1) Select cell B12.
(2) Type (B9)*0.134.
(3) Press ENTER.
(4) Select cell B14.
(5) Type (B9)*0.134*1440.
(6) Press ENTER.
(7) Select cell B16.
(8) Type (B9)*5.45.
(9) Press ENTER.

Enter the formulas for calculating distance from pumped well (r).

(1) Select cell G13.
(2) Type (G9)*(G9).
(3) Press ENTER.
(4) Select cell I9.
(5) Type (G9)*0.3048.
(6) Press ENTER.
(7) Select cell I13.
(8) Type (I9)*(I9).
(9) Press ENTER.

Enter the formulas for calculating the time–drawdown data.

(1) Select cell F25.
(2) Type ((H18*60) + C25)-((H18*60) + I18).
(3) Press ENTER.
(4) Select cell F26.
(5) Type ((B26*60) + C26)-((B25*60) + C25) + F25.
(6) Press ENTER.
(7) At cell F26, choose Edit | Copy.
(8) Select cell range of all time (min.) measurements (e.g., F27–F55).
(9) Choose Edit | Paste.
(10) Select cell G25.
(11) Type -(D25-D18).
(12) Select cell G26.

(13) Type -(D26-D18).

(14) Press ENTER.

(15) At cell G26, choose Edit | Copy.

(16) Select cell range of all drawdown (ft.) measurements (e.g., G27–G55).

(17) Choose Edit | Paste.

(18) Select cell H25.

(19) Type (F25) / 1440.

(20) Press ENTER.

(21) At cell H25, choose Edit | Copy.

(22) Select cell range of all time (days) measurements (e.g., H26–H55).

(23) Choose Edit | Paste.

(24) Select cell I25.

(25) Type (G25)*0.3048.

(26) Press ENTER.

(27) At cell I25, choose Edit | Copy.

(28) Select cell range of all drawdown (meters) measurements (e.g., I26–I55).

(29) Choose Edit | Paste.

Entering Field Measurements

Enter the name of the project.

(1) Select cell B4.

(2) Type the project name (e.g., Municipal Water Authority).

Enter the location of the project.

(1) Select cell B6.

(2) Type the location (e.g., Center Township).

Enter the pumping rate, in gpm.

(1) Select cell B9.

(2) Type the pumping rate (e.g., 200). The pumping rate in other measurement units will be calculated automatically and displayed in cells B12, B14, and B16.

Enter the static water level.

(1) Select cell D18.

(2) Type the static water level (e.g., 67.8).

Enter the identification code of the observation well.

(1) Select cell I4.

(2) Type the well identification code (e.g., No. 2).

Enter the identification code of the pumped well.

(1) Select cell I6.

(2) Type the well identification code (e.g., No. 1).

Enter the distance from pumped well to observation well.

(1) Select cell G9.

(2) Type the distance (e.g., 400). The distances, r (m) and r^2 (m²) will be calculated automatically and displayed in nearby cells.

Enter the time the pump was turned on (use 24-hour clock time).

(1) Select cell H18.

(2) Type the hour (e.g., 8).

(3) Select cell I18.

(4) Type the minute (e.g., 00).

Enter the date the pumping test began. (*Note:* format = "MM-DD-YR; be certain to type quotation marks before the numerals.)

(1) Select cell A25.

(2) Type the date.

Enter the time of first depth to water measurement. (You may ignore all pre-pumping measurements.)

(1) Select cell B25.

(2) Type the hour (e.g., 8).

(3) Select cell C25.

(4) Type the minute (e.g., 03).

Enter the first depth to water measurement, in feet.

(1) Select cell D25.

(2) Type the depth (e.g., 69.6).

Repeat the preceding procedure until all measurements of drawdown (and, if appropriate, recovery) have been entered on the spreadsheet template. *Caution:* If the clock time passes midnight (i.e., 2400 hr.), then be certain to add 24 to the hours reported in Column B. Time, t, and drawdown, s (as well as any recovery, s'), are calculated by the software program and displayed on the spreadsheet (Columns F through I) after each measurement is entered. Drawdown (and recovery) is displayed as a negative value. When all time–drawdown (and time–recovery) data pairs have been entered, compare the entered data with the original and correct any errors.

Modifying Table Format.

Delete all unnecessary rows of data.

(1) Select all rows of the table which do not contain measurement data (e.g., rows 51–54).

(2) Choose Edit|Cut.

Modify numeric format.

(1) Select measurement values in column D (e.g., cell range D25–D50).

(2) Change numeric format to one decimal place.

(3) Select drawdown values in column G (e.g., cell range G25–G50).

(4) Change numeric format to one decimal place.

(5) Select time values in column H (e.g., cell range H25–H50).

(6) Change numeric format to three decimal places.

(7) Select drawdown values in column I (e.g., cell range I25–I50).

(8) Change numeric format to two decimal places.

Saving the Time–Drawdown Data File

The time–drawdown data should be saved on disk for subsequent retrieval and printing. *Warning!* Do not save the data file under the name of the template file or the original template will be overwritten.

Save the new file for the first time.

(1) Select cell A1.

(2) Choose File|Save As. The Save File dialog box will appear on the screen. The name of an original file (e.g., fplot.txt) will be displayed and highlighted in the File Name text box.

(3) Press BACKSPACE to erase the name of the template text file.

(4) Type the new water level data file name (e.g., aquidat) in the File Name text box.

(5) Click on the arrow next to the File Type text box and select the type of file appropriate to your particular application. (Do not use *.txt.)

(6) Click on OK. The new file will be saved to the drive and directory you specified. Later, to save the existing file choose File|Save. The changes to the file will be saved under the original file name and the old data will be overwritten.

Printing the Time–Drawdown Data Form

Before you print, make sure that your application is properly configured for your particular printer. The instructions below provide only basic information about printing the table. For more detailed instructions, see the application reference manual or click on the Help icon.
Check the printer configuration.

(1) Choose File | Print. The Print dialog box will appear on the screen.

(2) Make sure that your printer is identified as active. (If not, use the printer setup option to select and configure it.)

Edit page settings.

(1) Click on the Page Setup button and, depending on your application, select:

For Quattro Pro: Print Scaling | Print to fit
For 1-2-3 in Size text box: Fit all to page
For Excel in Scaling box: Fit to: 1 page

(2) Click on OK.

Preview the print job.

(1) Click on the Print Preview button. A full-page view of the worksheet will be displayed. To view details, use the zoom option.

(2) Press ESC until the screen returns to the worksheet.

Print the table form.

(1) Choose File | Print. The Print dialog box will be displayed. The selected block appears in the Print or Print What box.

(2) Click on Print (or OK). The table form will be printed on your printer. If the format of the printout requires modifying, or if an error in data or labels is evident, return to the worksheet and make the necesary corrections. (The application user's guide or reference manual may be useful for these procedures.)

Closing the Spreadsheet Application and Quitting Windows

Before closing the application, be sure to save all active files.

(1) Choose File | Exit. If you have saved all active files, the application window closes. If you changed an active file but did not save it, then the Exit (or Save) dialog box will appear.

(2) If the Exit dialog box appears, choose Yes (and, if necessary, Replace) to save the file. The Windows desktop will appear on the screen.

(3) Open another application or quit Windows and turn off the computer. Be sure to close all open applications before quitting Windows.

CHAPTER 20

Method for Graphing
Time–Drawdown Curves

20.1 INTRODUCTION TO METHODS

THIS chapter describes three methods by which arithmetic time–drawdown (and recovery) curves can be graphed: the *Hand Plotting and Graphing Method* (Section 20.2), the *Microcomputer and Electronic Spreadsheet Method* (Section 20.3), and the *Microcomputer and Graphing Software Method* (Section 20.4).

Table 20.1 displays time–drawdown data for observation well No. 2 from the case study in Chapter 17. Use these data (or your own project data) for application of the methods described in this section.

20.2 HAND PLOTTING AND GRAPHING METHOD

Purpose

To construct a time–drawdown graph from aquifer test data.

Equipment and Materials

- aquifer test data (e.g., Table 20.1)
- arithmetic graph paper (e.g., Figure 20.1 or K & E 460780)
- 12-inch ruler/scale
- pencil (3H)
- French curve (optional)

Procedure

(1) Orient a sheet of arithmetic graph paper (e.g., Figure 20.1) so that the long dimension lies across the table in front of you.

TABLE 20.1. Time–Drawdown for Well No. 2
of the Case Study.

Time (min.)	Drawdown (ft.)	Time (min.)	Drawdown (ft.)
3	−1.8	100	−8.7
4	−2.3	120	−9.2
5	−2.6	150	−9.7
6	−2.9	180	−9.9
10	−3.9	240	−10.6
12	−4.3	300	−11.1
14	−4.6	360	−11.4
18	−5.1	480	−12.0
24	−5.7	600	−12.5
30	−6.2	720	−12.9
40	−6.7	960	−13.5
60	−7.6	1200	−14.0
80	−8.2	1440	−14.4

(2) Label the bottom axis "TIME (min.)" and the left axis "DRAWDOWN (ft.)."

(3) Lay out a drawdown scale on the left axis, with zero at the top horizontal line. Establish an axis scale such that the total drawdown fills as much of the axis as possible (see Figure 16.4 for an example).

(4) Lay out a time scale (minutes) on the bottom axis with zero at the left axis line. Establish a scale such that the total time of the test fills as much of the axis as possible (see Figure 16.4 for an example).

(5) Using the data from an aquifer test (e.g., Table 20.1), plot each time–drawdown data pair as a point on the graph. (If you have recovery data from an aquifer test, then you may plot each time–recovery level data pair also.)

(6) Complete the graph by connecting the time–drawdown data points with a smooth curve. (A French curve is helpful for drawing a smooth curve through the data points.)

20.3 MICROCOMPUTER AND ELECTRONIC SPREADSHEET METHOD

Purpose

To construct a time–drawdown graph from aquifer test data.

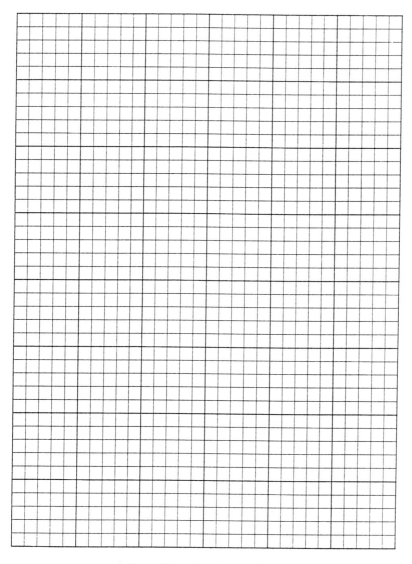

Figure 20.1 Arithmetic graph form.

Equipment and Materials

- aquifer test data (e.g., Table 20.1)
- microcomputer with an 80386 processor or higher, MS-DOS 3.3 or higher, Windows 3.1 or higher, hard drive, floppy disk drive, and graphics printer
- Windows version of electronic spreadsheet software (e.g., Lotus 1-2-3, Quattro Pro, or Excel)
- Hydrodata Diskette

Procedure

This method employs electronic spreadsheet software to construct a time–drawdown (and time–recovery) graph from aquifer test data (use Table 20.1 or your own project data). First, a data file of drawdown (and, if available, recovery) measurement levels at selected times is created and then a graph of these data is constructed. A copy of the completed drawdown graph can be produced on a printer or plotter.

Starting Up

Turn on the computer, monitor, and printer/plotter. Wait until the Windows desktop is displayed.

Open the application.

(1) Open the Windows group icon that contains your software application (e.g., Quattro Pro), or click on the Windows 95 Start button and point first to Programs and then to the application folder.

(2) Double click on the application icon (e.g., Quattro Pro) from the group window, or click on the name of the application from the drop-down list.

(3) Insert the Hydrodata Diskette into drive A (or B).

Entering Aquifer Test Data

The data file is constructed of two columns of data. The values of time, in minutes, are entered in column A of the worksheet, and the corresponding values of drawdown (and/or recovery) are entered in column B. Data entry begins with typing in the first time–drawdown data pair in the first row of cells (A1 and B1).

Enter the first time value, in minutes.

(1) Select cell A1.

(2) Type the time value (e.g., 3).

Enter the first drawdown value, in feet (use negative numbers for this procedure).

(1) Select cell B1.
(2) Type the drawdown value (e.g., −1.8).

Enter the second time value.

(1) Select cell A2.
(2) Type the time value (e.g., 4).

Enter the second drawdown value.

(1) Select cell B2.
(2) Type the drawdown value (e.g., −2.3).

Complete the data entry.

Repeat the data entry procedure until all time–drawdown pairs have been typed. If you have time–recovery data pairs, then enter these in sequence in the same columns directly following the time–drawdown data. After all the data have been typed in, compare the entries on the screen with the original data and correct all errors. When the data file is completed, save it to your data disk.

Saving the Time–Drawdown Data File

Save the new file for the first time.

(1) Select cell A1.
(2) Choose File | Save As. The Save As dialog box will appear on the screen.
(3) Select the File Name text box and type in the new file name (e.g., timedraw).
(4) Click on the arrow next to the File Type list box and select the type of file used by your particular software application.
(5) Specify the drive that contains the Hydrodata Diskette by clicking on the arrow next to the Drives list box and then clicking on the drive name (e.g., A).
(6) Click on OK. The new file will be saved to the drive and directory you specified. Later, to save the existing file chose File | Save.

Creating a Time–Drawdown (and/or Recovery) Graph

The specific procedures for creating and formatting a time–drawdown graph depend on the particular application software that you are using. In the following sections, refer to the instructions which apply to your particular application (e.g., Quattro Pro, 1-2-3, or Excel).

Name the graph/chart and specify the data to be plotted.

Quattro Pro

(1) Choose Graphics|New Graph. The New Graph dialog box will appear.

(2) With the Graph Name highlighted, type the name of your graph (e.g., TIMEDRAW). Do not press ENTER.

(3) Click on the X-Axis arrow. The New Graph dialog box will be minimized and the entire worksheet will be displayed again.

(4) Select (i.e., click and drag) the range of data (e.g., cells A1–A26) that represent the X-axis values.

(5) Press ENTER. The New Graph dialog box will reappear.

(6) Click on the 1st (series) arrow.

(7) Select the range of data (e.g., cells B1–B26) that represent the 1st (series) values.

(8) Press ENTER.

(9) If any information for other series is displayed, delete it and click on OK. The worksheet will disappear and a graph window displaying an initial version of the time–drawdown graph will appear on the screen.

(10) Go on to the next section (*Editing the Graph Type*).

1-2-3

(1) Select (i.e., click and drag) the range of data (e.g., cells A1–B26) to be graphed, including all headings and labels.

(2) Choose Tools|Chart. The cursor arrow is transformed into a chart pointer.

(3) Click and drag the chart pointer over a blank region of the worksheet (e.g., cells C3–H19). A chart illustrating an initial version of the time–drawdown graph is added to the worksheet. The Chart menu replaces the Range menu on the main menu bar.

(4) Choose Chart|Name.

(5) With the default chart name highlighted, type the name of your chart (e.g., TIMEDRAW).

(6) Click on Rename.

(7) Go on to the next section (*Editing the Graph Type*).

Excel

(1) Select (i.e., click and drag) the range of data (e.g., cells A1–B26) to be graphed, including all headings and labels.

(2) Choose Insert|Chart|On This Sheet. The cursor arrow is transformed into a chart pointer.

(3) Click and drag the chart pointer over a blank region of the worksheet (e.g., cells C3–H19). The ChartWizard dialog box Step 1 (of 5) is displayed.

(4) Choose Next>. The ChartWizard dialog box Step 2 (of 5) is displayed.

(5) Select the XY [Scatter] icon.

(6) Choose Next>. The ChartWizard dialog box Step 3 (of 5) is displayed.

(7) Select format icon #2 [data points and lines].

(8) Choose Next>. The ChartWizard dialog box Step 4 (of 5) and a sample chart is displayed.

(9) Choose Next>.

(10) On the Add a Legend edit line, select the No button.

(11) Select the Chart Title edit line.

(12) Type a chart title (e.g., TIME-DRAWDOWN - WELL NO. 2).

(13) In the Axis Titles text box, select the Category (X): text box.

(14) Type in the title of the X-axis [e.g., TIME (minutes)].

(15) Select the Value (Y): text box.

(16) Type in the title of the Y-axis [e.g., DRAWDOWN (feet)].

(17) Choose Finish. An initial version of the graph will be displayed on the worksheet.

(18) Go on to the next section (*Editing the Graph Type*).

Editing the Graph Type

If the displayed graph/chart is not an XY type, then proceed with the instructions in this section.

Quattro Pro

(1) Choose Graphics|Type. The Graph Types dialog box will be displayed.

(2) Select the XY icon.

(3) Choose OK.

(4) Go on to the next section (*Creating Graph Titles*).

1-2-3

(1) Choose Chart|Type.

(2) Select the XY Type button.

(3) Select the upper-left (data points and lines) icon.

(4) Click on OK.

(5) Go on to the next section (*Creating Graph Titles*).

Excel

The chart type was selected in a previous section. Go on to the next section (*Creating Graph Titles*).

Creating Graph Titles

Quattro pro

(1) Choose Graphics|Titles. The Graph Titles dialog box will be displayed.

(2) Select the Main Title edit line.

(3) Type in the main graph title (e.g., TIME-DRAWDOWN)

(4) Select the Sub Title edit line.

(5) Type in a subtitle (e.g., Well No. 2).

(6) Select the X-Axis Title edit line.

(7) Type in the X-axis title [e.g., TIME (minutes)].

(8) Select the Y-Axis Title edit line.

(9) Type in the Y-axis title [e.g., DRAWDOWN (feet)].

(10) Click on OK.

(11) Go on to the next section (*Editing the Graph Format*).

1-2-3

(1) Click anywhere inside the chart frame.

(2) Choose Chart|Headings.

(3) With the Title Line 1 edit line highlighted, type in the main chart title (e.g., TIME-DRAWDOWN).

(4) Select the Line 2 edit line.

(5) Type in a subtitle (e.g., Well No. 2).

(6) Click on OK.

(7) Go on the the next section (*Editing the Graph Format*).

Excel

A chart title was created in a previous section. Go on to the next section (*Editing the Graph Format*).

Editing the Graph Format

Edit the graph axes.

Quattro Pro

(1) Choose Property│Axis. The X-Axis dialog box will be displayed.
(2) Select Scale.
(3) Select the Increment edit line.
(4) Delete all text displayed on the line and type in new data increments (e.g., 200).
(5) Click on OK.
(6) Choose Property│Y Axis. The Y-Axis dialog box will be displayed.
(7) Select Numeric Format.
(8) Select the Fixed button.
(9) Click the arrow next to the Enter Number of Decimal Places edit line until the desired number of decimal places (e.g., 1) appear.
(10) Click on OK.
(11) Go on to the next section (*Editing the Data Legend*).

1-2-3

(1) Choose Chart│Axis│X-Axis. The X-Axis dialog box will be displayed.
(2) With the axis title highlighted, type in the title of the X-axis [e,g, TIME (minutes)].
(3) In the Scale Manually text box, select the Major Interval edit line.
(4) Delete any text displayed on the line and type in new major interval units for the axis (e.g., 200).
(5) In the Scale Manually text box, select the Minor Interval edit line.
(6) Delete any text displayed on the line and type in new minor interval units for the axis (e.g., 100).
(7) In the Show Tick Marks text box, select both the Major Interval and Minor Interval boxes.
(8) Click on OK.
(9) Choose Chart│Axis│Y-Axis. The Y-Axis dialog box will be displayed.

(10) With axis title selected (highlighted in blue), type in the name of the Y-axis [e.g., DRAWDOWN (feet)].

(11) In the Scale Manually text box, select the Major Interval edit line.

(12) Delete any text displayed on the line and type in new major interval units for the axis (e.g., 1.0).

(13) In the Scale Manually text box, select the Minor Interval edit line.

(14) Delete any text displayed on the line and type in new minor interval units (e.g., 0.5).

(15) In the Show Ticks Marks text box, select both the Major Interval and Minor Interval boxes.

(16) Click on OK.

(17) Go on to the next section (*Editing the Data Legend*).

Excel

(1) Double click anywhere inside the chart frame.

(2) Select the X-axis of the graph.

(3) Choose Format | Selected Axis. The Format Axis dialog box will be displayed.

(4) Choose the Patterns tab.

(5) In the Tick Mark Type text box, select the Major Outside button.

(6) Select the Minor Outside button.

(7) Choose the Scale tab.

(8) Select the Major Unit edit line.

(9) Delete any text displayed on the line and type in new major time units (e.g., 200).

(10) Select the Minor Unit edit line.

(11) Delete any text displayed on the line and type in new minor time units (e.g., 100).

(12) Click on OK.

(13) Click on the Y-axis of the graph.

(14) Choose Format | Selected Axis. The Format Axis dialog box will be displayed.

(15) Choose the Patterns tab.

(16) In the Tick Mark Type text box, select Major Outside.

(17) Select Minor Outside.

(18) Choose the Scale tab.

(19) Select the Major Unit edit line.

(20) Delete any text displayed on the line and type in new major drawdown units (e.g., 1.0).

(21) Select the Minor Unit edit line.

(22) Delete any text displayed on the line and type in new minor drawdown units (e.g., 0.5).

(23) Select Value (X) Axis, Crosses At: text box.

(24) Delete any text displayed on the line and type in a new X-axis origin (e.g., − 15.0).

(25) Click on OK.

(26) Go on to the next section (*Editing the Data Legend*).

Editing the Data Legend

Quattro Pro

No legend is displayed. Go on to the next section (*Editing Grid Lines*).

1-2-3

(1) Choose Chart|Legend. The Legend dialog box will be displayed.

(2) Select the Legend Entry edit line.

(3) Delete all text from the line. This action removes a data legend from the chart.

(4) Click on OK.

(5) Go on to the next section (*Editing Grid Lines*).

Excel

No legend is displayed on the chart. Go on to the next section (*Editing Grid Lines*).

Editing Grid Lines

Quattro Pro

(1) Choose Property|X Axis. The X-Axis dialog box will be displayed.

(2) Select Major Grid Style.

(3) Choose Line Style button.

(4) Select Solid Line Style box.

(5) Click on OK.

(6) Go on to the next section (*Saving the Graph*).

1-2-3

(1) Choose Chart|Grids. The Grids dialog box is displayed.

(2) Click on the arrow next to the X-Axis edit line.

(3) Select the desired grid-line interval (e.g., Major Interval).

(4) Click on the arrow next to the Y-Axis text box.

(5) Select the desired grid-line interval (e.g., Major Interval).

(6) Click on OK.

(7) Go on to the next section (*Saving the graph*).

Excel

(1) Choose Insert | Gridlines. The Gridlines dialog box will be displayed.

(2) Select the Major Gridlines boxes for both X and Y axes.

(3) Click on OK.

(4) Go on to the next section (*Saving the Graph*).

Saving the Time–Drawdown Graph and Data File

Save the graph/chart and data.

(1) Choose File | Save. The data and the graph will be saved under the original file name.

Printing/Plotting the Time–Drawdown Graph

Before you print/plot, make sure that your application is properly configured for your particular printer. The instructions below provide only basic information about printing/plotting the time–drawdown graph. For more detailed instructions, see the application reference manual or click on the Help icon.

Check the printer/plotter configuration.

(1) Choose File | Print. The Print dialog box will appear on the screen.

(2) Make sure your printer/plotter is identified as active. (If not, use the printer setup option to select and configure it.)

Preview the print job.

(1) Make sure that the graph/chart window (not the worksheet) is displayed and active.

(2) Click on the Print Preview button. A full-page view of the graph will be displayed. If desirable, modify the appearance of the page by using Page Setup.

(3) Press ESC until the screen returns to the graph.

Print/plot the time–drawdown graph.

(1) Choose File | Print. The Print dialog box will appear on the screen.

(2) Click on Print (or OK). The time–drawdown graph will be produced on the printer or plotter.

Saving a Text (ASCII) File of the Time–Drawdown Data

The columns of time–drawdown data which were created in this exercise may be useful in other applications, such as a scientific graphing program like GRAPHER. The exchange of data between different applications is facilitated if the data are in text (i.e., ASCII) form.

(1) Choose File|Save As. The Save File dialog box is displayed.
(2) Click on the arrow next to the File Types list box and select a name and extension that designates text files.
(3) Select the File Name text box.
(4) If necessary, delete all text in the box and type in a new file name (e.g., timedraw.txt).
(5) Click on OK. A dialog box will appear with a message warning you that graphs cannot be saved in the ASCII format; that is, only the data of the worksheet will be saved.
(6) Click on OK or Write.
(7) The time–drawdown data will be saved as a text (ASCII) file.

Closing the Spreadsheet Application and Quitting Windows

Before closing the application, be sure to save all active files.

(1) Choose File|Exit. If you have saved all active files, the application window closes. If you changed an active file but did not save it, then the Exit dialog box will appear.
(2) If the Exit dialog box appears, choose Yes to save the file. The Windows desktop will appear on the screen.
(3) Open another application or quit Windows and turn off the computer. Be sure to close all open applications before quitting Windows.

20.4 MICROCOMPUTER AND GRAPHING SOFTWARE METHOD

Purpose

To construct a time–drawdown graph from aquifer test data.

Equipment and Materials
- aquifer test data (e.g., Table 20.1)
- microcomputer with an 80386 processor or higher, MS-DOS 3.3 or higher, Windows 3.1 or higher, hard drive, floppy disk drive, and graphics printer

- GRAPHER for Windows (Version 1.25)
- Hydrodata Diskette

Procedure

This method employs GRAPHER for Windows to construct a time–drawdown (and time–recovery) graph from aquifer test data. The GRAPHER worksheet can be used to create and save a data file of time and drawdown (use Table 20.1 or your own project data). To enter time–drawdown data pairs from the keyboard, refer to *Entering Aquifer Test Data.* If you have already created and saved ASCII file of time–drawdown data pairs (e.g., timedraw.txt, which was created for the spreadsheet application in Section 20.3). [See *Opening a Text (ASCII) File of Aquifer Test Data.*]

Starting Up

Turn on the computer, monitor, and printer/plotter. Wait until the Windows desktop is displayed.

Open the application.

(1) Open the Windows group icon that contains your software application (e.g., Golden Software), or click on the Windows 95 Start button and point first to Programs and then to the application folder.

(2) Double click on the application icon (e.g., GRAPHER) from the group window, or click on the name of the application from the drop-down list. The GRAPHER Plot1 window will be displayed on the screen. (If the page frame appears in landscape orientation, you should reorient it to portrait. Choose File | Page Layout.)

(3) Insert the Hydrodata Diskette into drive A (or B).

Display the GRAPHER worksheet.

(1) Choose File | Worksheet. A worksheet window (labeled Sheet1) will appear superimposed over the plot window. If you have a text (ASCII) file of aquifer test data (e.g., filename: timedraw.txt), you may skip the next section of instructions and go on to *Opening a Text (ASCII) File of Aquifer Test Data.* If you must create a new data file for GRAPHER, then proceed with the following steps.

Entering Aquifer Test Data

The aquifer test data are arranged in two columns. The values of time, in minutes, are entered in column A of the worksheet, and the corresponding values of drawdown are entered in column B. Data entry begins with typing in the first time–drawdown data pair in the first row of cells (Al and Bl).

Enter the first time value, in minutes.

(1) Select cell A1.

(2) Type the time value (e.g., 3).

Enter the first drawdown value, in feet (use negative numbers).

(1) Select cell Bl.

(2) Type the drawdown value (e.g., − 1.8).

Enter the second time value.

(1) Select cell A2.

(2) Type the time value (e.g., 4).

Enter the second drawdown value.

(1) Select cell B2.

(2) Type the drawdown value (e.g., −2.3).

Complete the data entry. Repeat the data entry procedure until all time–drawdown pairs have been typed. If you have time–recovery data pairs, then enter these in sequence in the same columns directly following the time–drawdown data. After all the aquifer test data have been entered, compare the data entries on the screen with the original data and correct all errors. When the data file is completed, save it to your data disk.

Saving the Time–Drawdown Data File

Save the new file for the first time.

(1) Select cell A1.

(2) Choose File│Save As. The Save File dialog box will appear on the screen. The name of an original file (e.g., timedraw.txt) will be displayed and highlighted in the File Name text box.

(3) Press BACKSPACE to delete any displayed text and type the new water level data file name (e.g., timedraw) in the File Name text box.

(4) Click on the arrow next to the File Type list box and select ASCII files [*.DAT].

(5) Click on the arrow next to the Drives list box and select the drive that contains the Hydrodata Diskette (e.g., A or B).

(6) Click on OK. The new file will be saved to the drive and directory you specified. Later, to save the existing file choose File│Save. Skip the next section and go on to *Creating a Drawdown Graph*.

Opening a Text (ASCII) File of Aquifer Test Data

(1) If you have not inserted the Hydrodata Diskette into drive A (or B), do so now.

(2) Choose File | Worksheet.

(3) Choose File | Open. The Open Data dialog box will be displayed on the screen.

(4) Specify the drive that contains your file by clicking on the arrow next to the Drives list box and then clicking on the drive name (e.g., a:).

(5) Specify the file you want to open by selecting the name of the desired file in the File Name drop-down box (e.g., timedraw.txt).

(6) Click on OK. The water level data will be displayed in columns A and B.

Creating a Drawdown Graph

(1) Choose Window | Plot1.

(2) Choose Graph | Line or Symbol. The Select Worksheet dialog box will be displayed.

(3) Select the time–drawdown file (e.g., timedraw.txt).

(4) Click on OK. The Line Plot dialog box will be displayed. This box contains several specifications of the graph. (For more information, refer to the application user's manual.)

(5) Click on OK. The dialog box will disappear and an initial version of the hydrograph will be displayed.

Editing the Graph Format

Edit the X-axis.

(1) Select (i.e., click on) the X-axis of the graph.

(2) Choose Set | Axis. The Edit X Axis dialog box will be displayed.

(3) In the Length and Starting Position text box, click on the arrow next to the Length text box until it reads 5.00 in.

(4) In the Length and Starting Position text box, click on the arrow next to the X: text box until it reads 2.00 in.

(5) In the Length and Starting Position text box, click on the arrow next to the Y: text box until it reads 3.50 in.

(6) Select the Title edit box.

(7) Type in the title of the X-axis [e.g., TIME (minutes)].

(8) Choose the Edit Ticks button. The Tick Marks dialog box will be displayed.

(9) Select the Spacing (data units) edit line in the Major text box.

(10) Delete the default number (e.g., 400) and type in a new spacing value (e.g., 200).

(11) Click on OK.

(12) Choose the Edit Labels button. The Tick Labels dialog box will be displayed.

(13) Choose the Format button. The Label Format dialog box will appear.

(14) Click on the arrow next to the Decimal Digits edit box until a zero (0) is displayed.

(15) Click on the OK button of each dialog box until the plot window is displayed again.

Edit the Y-axis.

(1) Select the Y-axis of the graph.

(2) Choose Set│Axis. The Edit Y Axis dialog box will be displayed.

(3) In the Length and Starting Position text box, click on the arrow next to the Length: edit line until it reads 4.00 in.

(4) In the Length and Starting Position text box, click on the arrow next to the X: text box until it reads 2.00 in.

(5) In the Length and Starting Position text box, click on the arrow next to the Y: text box until it reads 3.50 in.

(6) Select the Title edit box.

(7) Type in the title of the Y-axis [e.g., DRAWDOWN (feet)].

(8) Choose the Edit Ticks button. The Tick marks dialog box will be displayed.

(9) Select the Spacing (data units) edit line in the Major text box.

(10) Delete the default number (e.g., 5) and type in a new spacing value (e.g., 1).

(11) Click on OK.

(12) Choose the Edit Labels button. The Tick Labels dialog box will be displayed.

(13) Choose the Format button. The Label Format dialog box will appear.

(14) Click on the arrow next to the Decimal Digits edit box until a one (1) is displayed.

(15) Click on the OK button of each dialog box until the plot window is displayed again.

(16) Select View│Zoom Page. This action will resize the time–drawdown graph so that the entire page is in view.

Edit the Line/Symbols.

(1) Select the line that marks the trend of the plotted data.

(2) Choose Set|Symbol Attributes. The Symbol Attributes dialog box will be displayed.

(3) Select the icon of the plot symbol you want to use (e.g., open circle).

(4) Click on OK.

Editing Grid Lines

(1) Select the X-axis.

(2) Choose Set|Grid Lines. The Grid Lines dialog box will be displayed.

(3) Select the At Major Ticks box.

(4) Click on OK.

(5) Select the Y-axis.

(6) Choose Set|Grid Lines. The Grid Lines dialog box will be displayed.

(7) Select the At Major Ticks box.

(8) Click on OK.

Adding a Graph Title

(1) Choose Draw|Text. The normal cursor arrow will change to an arrow with a T.

(2) Position the cursor somewhere near the page coordinates X = 2, Y = 9. (The exact location is not necessary because the graph title may be repositioned at a later time.)

(3) Click on the chosen location. The Text dialog box will be displayed.

(4) Click on the arrow next to the Points edit line until the default number (e.g., 12) is replaced by a new font size (e.g., 24).

(5) In the text box at the bottom of the dialog box, type in the graph title (e.g., TIME-DRAWDOWN - Well No. 2) in the space following the blinking cursor bar.

(6) Click on OK.

(7) On the time–drawdown graph, select (i.e., click on) the graph title.

(8) Click and drag the title to a suitable location on the page (e.g., centered approximately 1 inch above the top of the graph).

(9) Unselect the graph title by clicking once outside the plot frame.

Saving the Time–Drawdown Graph

Save the graph/chart and data.

(1) Choose File | Save As. The Save As dialog will be displayed.

(2) With the default file name (e.g., plotl.grf) highlighted, type in the new name for the time–drawdown graph (e.g., timedraw.grf).

(3) Click on OK. The data and time–drawdown graph will be saved under the original file name.

Printing the Time–Drawdown Graph

Before you print/plot, make sure that your application is properly configured for your particular printer. The instructions below provide only basic information about printing/plotting the graph. For more detailed instructions, see the application reference manual or click on the Help icon.

Check the printer/plotter configuration (optional).

(1) Choose File | Change Printer. The Change Printer dialog box will appear on the screen.

(2) Select your printer/plotter from the Printer list box. (Use Setup to specify print settings.)

(3) Click on OK.

Print/plot the graph.

(1) Make sure that the graph (not the worksheet) is displayed and active.

(2) Choose File | Print.

(3) Click on OK.

Closing the Application and Quitting Windows

Before closing the application, be sure to save all active files.

(1) Choose File | Exit. If you have saved all active files, the application window closes. If you changed an active file but did not save it, then the Exit dialog box will appear.

(2) If the Exit dialog box appears, choose Yes to save the file. The Windows desktop will appear on the screen.

(3) Opan another application or quit Windows and turn off the computer. Be sure to close all open applications before quitting Windows.

CHAPTER 21

Analytical Procedure for Estimating Hydraulic Properties of Ideal Aquifers: Theis Nonequilibrium Method

21.1 INTRODUCTION TO METHODS

THEIS (1935) developed a graphical solution to the nonequilibrium equation that yields values of transmissivity and storativity if other terms are known (see Chapter 16 for a background discussion). The method involves matching a data plot of time and drawdown on logarithmic graph paper with a Theis type curve, which is prepared by plotting values of $W(u)$ against $1/u$ on logarithmic graph paper (see Figure 21.1). If the assumptions that underlie the Theis equation are satisfied, then the type curve has the same general shape as a profile of the cone of depression, and, thus, the drawdown graph at any point in the cone of depression.

In this procedure, time–drawdown data are plotted on logarithmic graph paper on which time since pumping began defines the horizontal axis and drawdown defines the vertical axis (see Figure 21.2). The type curve graph is then superimposed on the data plot in such a way that the plotted points fall on the type curve (see Figure 21.3). After a close correspondence between the data points and the type curve is found, a match point is selected. While any convenient point on the graphs may be chosen, subsequent calculations are made easier if the match point is where $1/u$ equals, 1, 10, or 100 and $W(u)$ equals an integer (see Figure 21.3). The example in Figure 21.3 illustrates a match point at $s = 0.58$ feet, $W(u) = 1.0$, $t = 89$ minutes, and $1/u = 100$.

Transmissivity is calculated from the equation

$$T = \frac{QW(u)}{4\pi s} \tag{21.1}$$

241

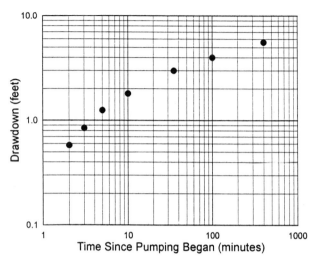

Figure 21.1 Aquifer test data plotted on logarithmic graph paper (modified from Driscoll, 1986, p. 264).

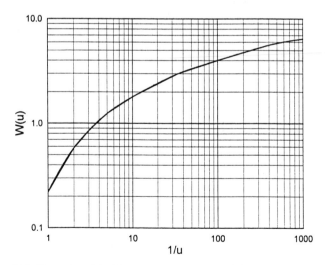

Figure 21.2 Illustration of the Theis type curve (modified from Driscoll, 1986, p. 264).

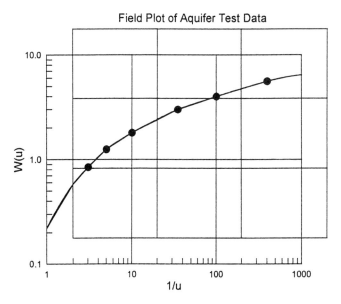

Figure 21.3 Aquifer test data superimposed on the Theis type curve (modified from Driscoll, 1986, p. 264).

where T is the transmissivity $[L^2/T]$, Q is the pumping rate $[L^3/T]$, s is the drawdown $[L]$, $W(u)$ is the "well function" of u, and

$$u = \frac{r^2 S}{4Tt} \qquad (21.2)$$

where r is the distance $[L]$ from pumped well to observation well, S is the storativity [dimensionless], defined below, and t is the time since pumping started $[T]$. Storativity is calculated from the equation

$$S = \frac{4Ttu}{r^2} \qquad (21.3)$$

where S is the storativity (dimensionless), T is the transmissivity $[L^2/T]$, t is the time since pumping started $[T]$, and r is the distance $[L]$ from pumped well to observation well. These equations require the use of consistent measurement units. For example, if transmissivity is desired in units of ft.^{2}/min., then pumping rate must be in ft.^{3}/min., drawdown and distance in ft., and time in min. In the metric system the common units are, respectively, m^2/sec., m^3/sec., m, and sec.

TABLE 21.1. Aquifer Test Data for Well No. 2 of the
Case Study (Q = 200 gpm,[1] r = 400 ft.).

Drawdown		Drawdown	
Time (min.)	Drawdown (ft.)	Time (min.)	Drawdown (ft.)
3	1.8	100	8.7
4	2.3	120	9.2
5	2.6	150	9.7
6	2.9	180	9.9
10	3.9	240	10.6
12	4.3	300	11.1
14	4.6	360	11.4
18	5.1	480	12.0
24	5.7	600	12.5
30	6.2	720	12.9
40	6.7	960	13.5
60	7.6	1200	14.0
80	8.2	1440	14.4

[1]Before beginning calculations and plotting, convert Q from gpm to ft.^{3}/min.

This requirement causes a problem if American practical hydrology units are involved in the procedure. That is, for example, where pumping rate is measured and reported in gpm. In this case, two unit conversions are necessary. Before entering a value of Q into the equation, convert gpm to ft.^{3}/min. by the following formula: ft.^{3}/min. = gpm $\times$ 0.134. [For additional information, see Driscoll (1986), pp. 260–265, and Heath (1987), pp. 36–37.]

This chapter describes two methods by which Theis type curves can be constructed and aquifer coefficients calculated. They are the *Hand Plotting and Curve Matching Method* (Section 21.2) and the *Microcomputer and Type Curve Matching Software Method* (Section 21.3).

Table 21.1 displays aquifer test data from the case study in Chapter 17. Use these data (or your own project data) for application of the methods described in this chapter.

21.2 HAND PLOTTING AND CURVE MATCHING METHOD

Purpose

To estimate the transmissivity and storativity of an ideal aquifer by the Theis type curve match method.

References

Driscoll, F. G., 1986. *Groundwater and Wells.* (2nd ed.) St. Paul, MN: Johnson Division, pp. 260–265.

Heath, R. C., 1987. *Basic Ground-Water Hydrology.* U.S . Geological Survey Water-Supply Paper 2220, pp. 36–37.

Equipment and Materials

- aquifer test data (e.g., Table 21.1)
- logarithmic graph paper (e.g., Figure 21.4)
- Theis type curve (e.g., Figure 21.5)
- Theis curve match worksheet (see Figure 21.6)
- light table (optional)
- pencil (3H)

Procedure

Label field plot graph axes. On a sheet of logarithmic graph paper (e.g., Figure 21.4), label the bottom (horizontal) axis "TIME SINCE PUMP STARTED (min)" and the left (vertical) axis "DRAWDOWN (ft)." See Figure 21.2 for an example. (If necessary, renumber the first log cycle of the vertical (left) axis 0.1, 0.2, 0.3, etc., beginning at the bottom with 0.1. Also, renumber the second and third log cycles 10, 20, 30, etc., and then 100, 200, 300, etc.). See Figure 21.2 for an example.

Plot time–drawdown data. Using aquifer test data (e.g., Table 21.1, or your own project data), plot time–drawdown data pairs on the log-log graph paper (e.g., Figure 21.4).

Superimpose the Theis curve graph on plot of field data. On a light table (or, in lieu of this, a windowpane) superimpose the Theis type curve (Figure 21.5) on the plot of time–drawdown data. Alternatively, you may use a Theis type curve transparency to overlay the field data plot.

Select a match point. Keeping the axes of both graphs in parallel alignment, select a match point for the data plot and the type curve. For example, select the point on the Theis curve graph with the coordinates $1/u = 100$ and $W(u) = 1$. Then, read the coordinates of t and s off the data plot graph that correspond to the selected match point. Record the match point values s, $W(u)$, t, and $1/u$ under MATCH POINTS section of the Theis curve match worksheet (Figure 21.6).

Calculate transmissivity. Enter the values of Q, t, $W(u)$, and s on the corresponding lines under the TRANSMISSIVITY section of the Theis curve match worksheet. (Be sure to convert gpm to ft.^{3}/min.) Use the equation to calculate the transmissivity of the aquifer and record this value on the appropriate line of the Theis curve match worksheet.

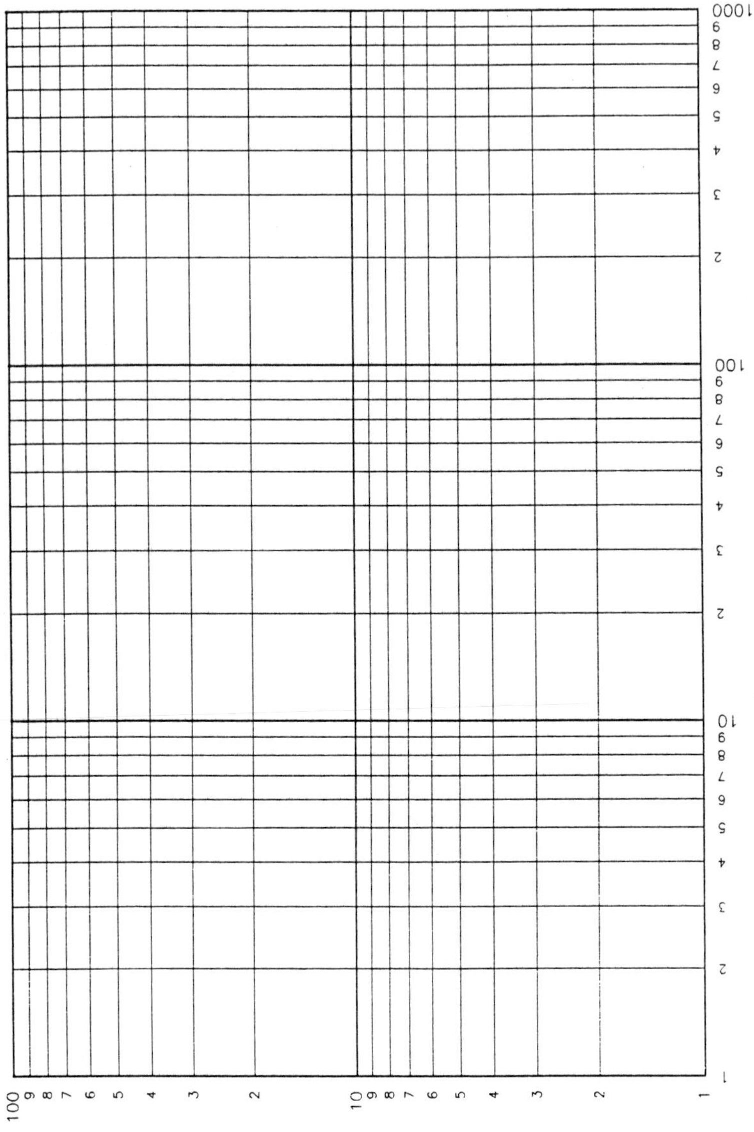

Figure 21.4 Logarithmic graph form.

246

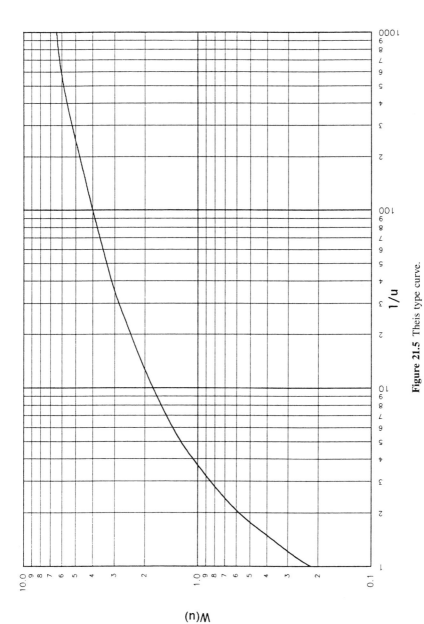

Figure 21.5 Theis type curve.

247

THEIS PLOT MATCH POINTS

Data Plot Values

$$s \quad = \quad \underline{\hspace{2cm}} \quad ft$$

$$t \quad = \quad \underline{\hspace{2cm}} \quad min$$

Type Curve Values

$$W(u) \quad = \quad \underline{\hspace{2cm}}$$

$$1/u \quad = \quad \underline{\hspace{2cm}}$$

TRANSMISSIVITY

$$Q \quad = \quad \underline{\hspace{2cm}} \quad gpm \quad = \quad \underline{\hspace{2cm}} \quad ft^3/min$$

$$t \quad = \quad \underline{\hspace{2cm}} \quad min$$

$$W(u) \quad = \quad \underline{\hspace{2cm}}$$

$$s \quad = \quad \underline{\hspace{2cm}} \quad ft$$

$$T \quad = \quad \frac{QW(u)}{4\pi s} \quad = \quad \underline{\hspace{2cm}} \quad ft^2/min$$

STORATIVITY

$$u \quad = \quad \underline{\hspace{2cm}}$$

$$r \quad = \quad \underline{\hspace{2cm}} \quad ft$$

$$r^2 \quad = \quad \underline{\hspace{2cm}} \quad ft^2$$

$$S \quad = \quad \frac{4Ttu}{r^2} \quad = \quad \underline{\hspace{2cm}}$$

Figure 21.6 Theis curve match worksheet.

248

Calculate storativity. Enter the values of u and r on the appropriate lines on the worksheet and calculate r^2. Use the equation to calculate the storativity of the aquifer and record this value on the Theis curve match worksheet.

21.3 MICROCOMPUTER AND TYPE CURVE MATCHING SOFTWARE METHOD

Purpose

To estimate the transmissivity and storativity of an ideal aquifer by the Theis type curve match method.

Reference

Duffield, G. M., and Rumbaugh, J. O., III, 1989. *AQTESOLV: Aquifer Test Solver, Version 1.00, Documentation* (Version 1.10 update). Geraghty & Miller, Inc.

Equipment and Materials

- aquifer test data (e.g., Table 21.1)
- microcomputer with an 80386 processor or higher, MS-DOS 3.3 or higher, Windows 3.1 or higher, hard drive, floppy disk drive, and HP laser printer or plotter.
- AQTESOLV for DOS (Version 1.10) software
- Hydrodata Diskette

Procedure

This method employs AQTESOLV to estimate aquifer properties and to construct a Theis graph from pumping test data. The program can be used to create and save a file of aquifer test data (see Table 21.1 for an example, or use your own project data). Alternatively, if you have an existing AQTE-SOLV file of aquifer test data (e.g., ideal.dat), it may be imported and used to construct a Theis graph.

Starting Up

AQTESOLV is DOS-based and can be run either from the DOS prompt (C:\>) or from the Windows *Run* command. (The Path command in your AUTOEXEC.BAT file should include AQTESOLV.)

(1) Turn on the computer, monitor, and plotter/printer. Wait until the Windows desktop is displayed.
(2) Choose File | Run, or click on the Windows 95 Start button and choose Run. The Run dialog box will appear.
(3) In the edit box, type AQTESOLV.
(4) Press ENTER. The AQTESOLV Program Control menu will be displayed on the screen.
(5) Insert the Hydrodata Diskette into drive A (or B).

Specify disk drive and directory to be used to retrieve and save data files.

(1) Choose Misc. Utilities.
(2) Choose Change Current Working Directory. The Change Current Working Directory dialog box will be displayed.
(3) Type A:\ (or B:\).
(4) Press ENTER ESC.

Creating an Aquifer Test Data File

If you are going to use a previously created AQTESOLV file of aquifer test data, you may skip this section of instructions and go on to *Loading an Aquifer Test Data File.*
Initialize a new data set.

(1) Choose Data Set Manager.
(2) Choose Create New Data Set. The Input Options menu will be displayed on the screen. *Message:* Are you sure you want to initialize the data set? (Y/N): No.
(3) Type Y ENTER.
(4) Choose Title for Data Set. *Message:* Enter title.
(5) Type the data set title (e.g., IDEAL AQUIFER ANALYSIS).
(6) Press ENTER.

Specify the aquifer test data.

(1) Choose Confined/Leaky Aquifer Test Data. The Confined/Leaky Aquifer Test Data menu will be displayed on the screen.
(2) Type the pumping rate (e.g., 26.8).
(3) Press ENTER.
(4) Type the distance to observation well (e.g., 400).
(5) Press ENTER ESC. The Confined/Leaky Aquifer Test Data menu will disappear from the screen.

(6) Choose Observation Well Measurements. The Time–Drawdown Entry menu will be displayed on the screen.

(7) Choose Enter Data from Keyboard. The observation well data worksheet will be displayed on the screen.

(8) Type the first time measurement (e.g., 3).

(9) Press ENTER.

(10) Type the first drawdown measurement (e.g., 1.8).

(11) Press ENTER ENTER.

(12) Type the second time measurement (e.g., 4).

(13) Press ENTER.

(14) Type the second drawdown measurement (e.g., 2.3).

(15) Press ENTER ENTER. Repeat this procedure until alll time–drawdown pairs have been entered. Check the entry data against the original measurements and correct any errors.

(16) Press ESC ESC ESC. *Message:* Save AQTESOLV data set on disk? (Y/N): Yes.

Saving the Aquifer Test Data File

(1) Press ENTER. *Message:* Enter file name for data set:

(2) Type the new data set filename (e.g., ideal.dat).

(3) Press ENTER. *Message:* Open new output file on disk? (Y/N): Yes.

(4) Press ENTER. *Message:* Enter output filename:

(5) Type the new output filename (e.g., theis.out).

(6) Press ENTER ESC.

Skip the next section and go on to *Analyzing the Data Set.*

Loading an Aquifer Test Data File

(1) Choose Data Set Manager.

(2) Choose Read Data Set from Disk. *Message:* Enter data set directory:

(3) Type the disk drive\directory (e.g., A:\ or B:\).

(4) Press ENTER. *Message:* Enter data set extension: *.dat

(5) Press ENTER. *Message:* Enter data set filename:

(6) Type the data set filename (e.g., ideal.dat).

(7) Press ENTER. The time–drawdown data set will be displayed on the screen.

(8) Press ENTER ENTER. *Message:* Open new output file on disk? (Y/N): Yes.

(9) Press ENTER.

(10) Type the output filename (e.g., THEIS.OUT).

(11) Press ENTER ESC. The Program Control menu will be displayed on the monitor screen.

Analyzing the Data Set.

(1) Choose Confined Solutions. The Confined Solutions menu will be displayed on the screen.

(2) Choose 1 Theis Method. The Parameter Estimation Control menu will be displayed on the screen.

(3) Choose Estimate Aquifer Parameters. The program will automatically adjust aquifer coefficients to match the Theis type curve to your time–drawdown data and display the values. *Message:* Perform additional nonlinear iterations? (Y/N): Yes.

(4) Type N ENTER.

(5) Choose Graph Results | Plot Graph. A Theis type curve and field plot data will be displayed on the screen. The estimates of transmissivity, T, and storativity, S, are printed in the upper-left corner of the graph. If time values have been entered in minutes, then transmissivity will be calculated in ft.^{2}/min.

Adjusting the Type Curve (optional)

By moving the mouse you may change the position of the type curve on the screen in order to obtain the "best" match with the data points. The values of T and S will change automatically to reflect the new curve match. After a match has been achieved, refresh the screen and replot the graph with the new curve match by pressing F4.

Printing/Plotting the Theis Graph

The instructions below assume that AQTESOLV is configured correctly for your particular printer or plotter. (If you need assistance, see the AQTESOLV documentation manual for configuring your hardware correctly.) You can output the Theis graph to either an HP plotter or a printer. Choose the appropriate option below.

Options (select one)

Plot graph on an HP plotter.

(1) Place a black pen in pen position #1 and insert a sheet of blank paper.

(2) Press ALT-P. *Message:* Is your plotter connected? (Y/N): Yes.

(3) Press ENTER.

Print graph on a printer.

(1) Press ALT-L. *Message:* Sending plot to a printer. Strike a key (ESC = abort).

(2) Press ENTER. After a short wait, a Theis type curve will be plotted/printed.

Saving the Theis Graph on Disk

(1) Press ESC ESC.

(2) Choose Print Results to Disk. The computer will beep twice when the graph has been saved on the disk.

Quitting the AQTESOLV Program

(1) Press F2 or choose Quit AQTESOLV from the Program Control menu. *Message:* Quit AQTESOLV? (Y/N): No.

(2) Type Y ENTER. The AQTESOLV program will disappear from the screen. If you ran the program from Windows, the Windows desktop will appear on the screen.

(3) Open another application or quit Windows and turn off the computer. Be sure to close all open applications before quitting Windows.

21.4 REFERENCES

Driscoll, F. G., 1986. *Groundwater and Wells* (2nd ed.). St. Paul, MN: Johnson Division.

Duffield, G. M., and Rumbaugh, J. O., III, 1989. *AQTESOLV; Aquifer Test Solver, Version 1.00, Documentation* (Version 1.10 update). Geraghty & Miller, Inc.

Heath, R. C., 1987. *Basic Ground-Water Hydrology.* U.S. Geological Survey Water-Supply Paper 2220.

Theis, C. V. 1935. The relation between the lowering of the piezometric surface and the rate and duration of discharge of a well using ground water storage. *Am. Geophysical Union Trans.*, pp. 518–525.

Analytical Procedure for Estimating Hydraulic Properties of Ideal Aquifers: Modified Theis Nonequilibrium (Cooper-Jacob) Method

22.1 INTRODUCTION TO METHODS

THE Cooper-Jacob Method permits an approximate solution to the Theis nonequilibrium equation using a straight-line (semi-logarithmic) graphical approach. If other terms, including pumping rate and distance from pumping well to observation wells are known, it is possible to estimate transmissivity and storativity of the aquifer from values read from the semi-logarithmic graph (see Figure 22.1). Heath (1987, p. 38) offers an important cautionary note: "However, it is essential to note that, whereas the Theis equation applies at all times and places (if the assumptions are met), Jacob's method applies only under certain additional conditions. These conditions must also be satisfied in order to obtain reliable answers." The chief condition is that the method is applicable only after the cone of depression has achieved a steady shape. This condition is met only when u is very small, i.e., 0.005 or less.

Transmissivity is calculated from the equation

$$T = \frac{2.3Q}{4\pi\Delta s} \tag{22.1}$$

where T is the transmissivity $[L^2/T]$, Q is the pumping rate $[L^3/T]$, and Δs is the change in drawdown $[L]$ across one log cycle. Storativity (coefficient of storage) is calculated from the equation

$$S = \frac{2.25 T t_0}{r^2} \tag{22.2}$$

where S is the storativity [dimensionless], T is the transmissivity $[L^2/T]$,

255

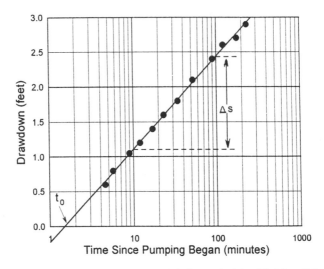

Figure 22.1 Illustration of the Cooper-Jacob straight-line method (modified from Driscoll, 1986, p. 264).

t_0 is the intercept of the straight line at zero drawdown, and r is the distance [L] from the pumped well to the observation well.

These equations require the use of consistent measurement units; for example, transmissivity in ft.^{2}/min., pumping rate in ft.^{3}/min., and drawdown and distance in ft. (Older publications may report transmissivity in gpd/ft., but this form is outmoded and is not recommended.) In the metric system the units would be, respectively, m^2/min., m^3/min., and m. [For additional information, see Driscoll (1986), pp. 219–223, and Heath (1987), pp. 38–39.]

This chapter describes three methods by which Cooper-Jacob (semilogarithmic) graphs for nonleaky artesian systems can be constructed and aquifer coefficients T and S can be estimated. These methods are: the *Hand Plotting and Graphing Method* (Section 22.2), the *Microcomputer and Graphing Software Method* (Section 22.3), and the *Microcomputer and Type Curve Matching Software Method* (Section 22.4).

Table 21.1 displays the aquifer test from the case study in Chapter 17. Use the drawdown data (or your own project data) for application of the methods described in this section.

22.2 HAND PLOTTING AND GRAPHING METHOD

Purpose

To construct a Cooper-Jacob graph and estimate the transmissivity and storativity of an aquifer.

References

Driscoll, F. G., 1986. *Groundwater and Wells.* (2nd ed.) St. Paul, MN: Johnson Division, pp. 219–223.

Heath, R. C., 1987. *Basic Ground-Water Hydrology.* U.S. Geological Survey Water-Supply Paper 2220, pp. 38–39.

Equipment and Materials

- aquifer test data (e.g., Table 21.1)
- logarithmic graph paper (e.g., Figure 22.2 or K & E 466213)
- Cooper-Jacob plot worksheet (see Figure 22.3)
- 12-inch ruler/scale
- pencil (3H)

Procedure

Label the graph axes.

(1) Orient a sheet of semi-logarithmic graph paper (e.g., Figure 22.2) so that the logarithmic axis lies perpendicular across the table in front of you.

(2) Label the log (bottom horizontal) axis "TIME (min)" and the arithmetic (left vertical) axis "DRAWDOWN (ft)."

Scale the graph axes.

(1) First, lay out the drawdown scale on the arithmetic (left vertical) axis with zero at the bottom horizontal line. Establish the scale divisions such that the maximum drawdown fills as much of the axis as possible (see Figure 22.1 for an example).

(2) Then, lay out the time scale on the logarithmic axis beginning with 1 and ending with 1000 (i.e., 1,2,3, . . . 10, 20, 30, . . . 100, 200, 300, etc.). Establish a scale such that the maximum time fills as much of the axis as possible (see Figure 22.1 for an example).

Plot time–drawdown data. Using aquifer test data (e.g., Table 21.1) plot each time–drawdown data pair as a point on the graph.

Construct a "best-fit" line.

(1) Complete the plot by drawing a straight line through the data points. Ignore the data points that represent measurements made during the first few minutes of the test and which do not fall on the straight line.

(2) Extend the straight line to the zero drawdown line of the graph and mark the value of the time intercept (t_0); see Figure 22.1 for an example.

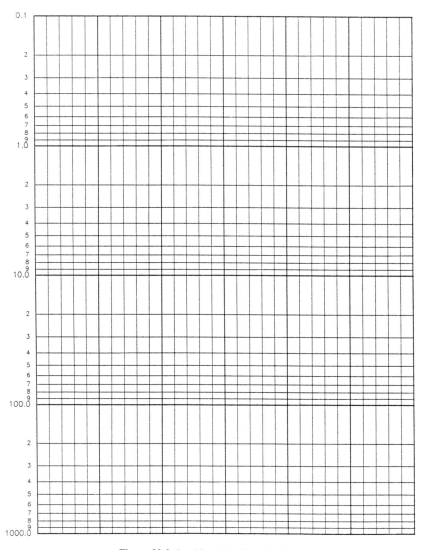

Figure 22.2 Semi-logarithmic graph form.

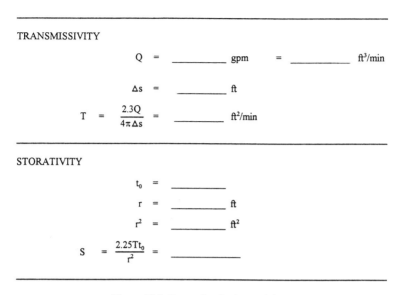

Figure 22.3 Cooper-Jacob plot worksheet.

Determine change in drawdown. Determine the change in drawdown (Δs) across one log cycle and mark it on the graph (see Figure 22.1 for an example).

Calculate transmissivity.

(1) Enter the values of Q and Δs on the appropriate lines on the Cooper-Jacob plot worksheet (Figure 22.3).

(2) Calculate T and record the value of the Cooper-Jacob plot worksheet.

Calculate storativity.

(1) Enter the values of t_0 and r on the appropriate lines of the worksheet.

(2) Calculate S and record this value on the Cooper-Jacob plot worksheet.

22.3 MICROCOMPUTER AND GRAPHING SOFTWARE METHOD

Purpose

To construct a Cooper-Jacob graph and estimate the transmissivity and storativity of an aquifer.

Equipment and Materials

• aquifer test data (e.g., Table 21.1)

- Cooper-Jacob plot worksheet (e.g., Figure 22.3)
- microcomputer with an 80386 processor or higher, MS-DOS 3.3 or higher, Windows 3.1 or higher, hard drive, floppy disk drive, and graphics printer or plotter
- GRAPHER for Windows (Version 1.25)
- Hydrodata Diskette

Procedure

This method employs GRAPHER for Windows to construct a Cooper-Jacob time–drawdown graph from aquifer test data. The GRAPHER worksheet can be used to create and save a file of aquifer test data (use Table 21.1 or your own project data). If you are entering data from the keyboard, refer to *Entering Time–Drawdown Data*. Alternatively, an existing text (ASCII) file of aquifer test data (e.g., ideal.dat)—use positive values only—may be imported and used to construct a Cooper-Jacob graph. [See *Opening a Text (ASCII) File of Aquifer Test Data.*]

Starting Up

Turn on the computer, monitor, and printer/plotter. Wait until the Windows desktop is displayed.
Open the application.

(1) Open the Windows group icon that contains your software application (e.g., Golden Software), or click on the Windows 95 Start button and point first to Programs and then to the application folder.

(2) Double click on the application icon (e.g., GRAPHER) from the group window, or click on the name of the application from the drop-down list. The GRAPHER Plot1 window will be displayed. (If the page frame appears in landscape orientation, you should reorient it to portrait. Choose File|Page Layout. Select Portrait and click on OK.)

(3) Insert the Hydrodata Diskette into drive A (or B).

Display the GRAPHER worksheet.

(1) Choose File|Worksheet. A worksheet window (labeled Sheet1) will appear superimposed over the plot window. If you have a text (ASCII) file of aquifer test data (e.g., filename: ideal.dat), you may skip the next section of instructions and go on to *Opening a Text (ASCII) File of Aquifer Test Data*. If you must create a new data file for GRAPHER, then proceed with the following steps.

Entering Aquifer Test Data

The aquifer test data are arranged in two columns. The values of time, in minutes, are entered in column A of the worksheet, and the corresponding values of drawdown are entered in column B. Data entry begins with typing in the first time–drawdown data pair in the first row of cells (A1 and B1).

Enter the first time value, in minutes.

(1) Select cell A1.

(2) Type the time value (e.g., 3).

Enter the first drawdown value, in feet (use positive numbers).

(1) Select cell B1.

(2) Type the drawdown value (e.g., 1.8).

Enter the second time value.

(1) Select cell A2.

(2) Type the time value (e.g., 4).

Enter the second drawdown value.

(1) Select cell B2.

(2) Type the drawdown value (e.g., 2.3)

Complete the data entry. Repeat the data entry procedure until all time–drawdown pairs have been typed. After all the aquifer test data have been typed in, compare the data entries on the screen with the original data and correct all errors. When the data file is completed, save it to your data disk.

Saving the Aquifer Test Data File

Save the new file for the first time.

(1) Select cell A1.

(2) Choose File | Save As. The Save As dialog box will appear on the screen.

(3) Press BACKSPACE to delete any displayed text and type in the new file name (e.g., idealcj) in the File Name text box.

(4) Click on the arrow next to the File Type text box and select ASCII files (*.DAT).

(5) Click on the arrow next to the Drives list box and select the drive that contains the Hydrodata Diskette (e.g., A or B).

(6) Click on OK. The new file will be saved to the drive and directory you specified. Later, to save the existing file choose File | Save. The changes to the file will be saved under the original file name and the old data will be overwritten. Skip the next section and go on to *Creating a Cooper-Jacob Graph.*

Opening a Text (ASCII) File of Aquifer Test Data

(1) If you have not inserted the Hydrodata Diskette into drive A (or B), do so now.
(2) Choose File | Open. The Open Data dialog box will be displayed on the screen.
(3) Click on the arrow next to the File Type list box and select ASCII data (*.txt) from the drop-down list.
(4) Specify the drive that contains your file by clicking on the arrow next to the drives list box and then double clicking on the drive name (e.g., A).
(5) Specify the file you want to open by selecting the name of the desired file in the file name drop-down box (e.g., ideal.dat).
(6) Click on OK. The aquifer test data will be displayed in the worksheet window. (If you are using a data file created by AQTESOLV, GRAPHER will ignore the data in lines 1–6.)

Creating a Cooper-Jacob Graph

(1) Choose Window | Plot1.
(2) Choose Graph | Line or Symbol. The Select Worksheet dialog box will be displayed.
(3) Select the time–drawdown file (e.g., ideal.dat).
(4) Click on OK. The Line Plot dialog box will be displayed. This box contains several specifications of the graph. (For more information, refer to the application user's manual.)
(5) Click on OK. The dialog box will disappear and an initial version of the Cooper-Jacob graph will be displayed.

Editing the Graph Format

Edit the X-axis
(1) Select (i.e., click on) the X-axis of the graph.
(2) Choose Set | Axis. The Edit X Axis dialog box will be displayed.

(3) In the Length and Starting Position text box, click on the arrow next to the Length text box until it reads 5.00 in.

(4) In the Length and Starting Position text box, click on the arrow next to the X: text box until it reads 2.00 in.

(5) In the Length and Starting Position text box, click on the arrow next to the Y: text box until it reads 3.50 in.

(6) Click on the arrow next to the Scale edit box and choose Log (base 10).

(7) In the Axis Limits text box, select the Axis Min. text box, delete any default value and type a new value (e.g., 1).

(8) Select the Title edit box.

(9) Type in the title of the X-axis [e.g., TIME (minutes)].

(10) Choose the Edit Labels button. The Tick Labels dialog box will be displayed.

(11) Choose the Format button. The Label Format dialog box will appear.

(12) Click on the arrow next to the Decimal Digits edit box until a zero (0) is displayed.

(13) Click on the OK button of each dialog box until the plot window is displayed again.

Edit the Y-axis.

(1) Select the Y-axis of the graph.

(2) Choose Set|Axis. The Edit Y Axis dialog box will be displayed.

(3) In the Length and Starting Position text box, click on the arrow next to the Length: edit line until it reads 4.00 in.

(4) In the Length and Starting Position text box, click on the arrow next to the X: text box until it reads 2.00 in.

(5) In the Length and Starting Position text box, click on the arrow next to the Y: text box until it reads 3.50 in.

(6) Select the Title edit box.

(7) Type in the title of the Y-axis [e,g., DRAWDOWN (feet)].

(8) Choose the Edit Ticks button. The Tick Marks dialog box will be displayed.

(9) Select the Spacing (data units) edit line in the Major text box.

(10) Delete the default number (e.g., 5) and type in a new spacing value (e.g., 1).

(11) Click on OK.

(12) Choose the Edit Labels button. The Tick Labels dialog box will be displayed.

(13) Choose the Format button. The Label Format dialog box will appear.
(14) Click on the arrow next to the Decimal Digits edit box until a one (1) is displayed.
(15) Click on the OK button of each dialog box until the plot window is displayed again.
(16) Select View | Zoom Page. This action will resize the graph so that the entire page is in view.

Edit the Line/Symbols.

(1) Select the line that marks the trend of the plotted data.
(2) Choose Set | Symbol Attributes. The Symbol Attributes dialog box will be displayed.
(3) Select the icon of the plot symbol you want to use (e.g., open circle).
(4) Click on OK.
(5) Choose Set | Line Attributes. The Line Attributes dialog box will be displayed.
(6) In the Style edit box, select invisible (i.e., blank) symbol.
(7) Click on OK.

Editing Grid Lines

(1) Select the X-axis.
(2) Choose Set | Grid Lines. The Grid Lines dialog box will be displayed.
(3) Select the At Major Ticks box.
(4) Click on OK.
(5) Select the Y-axis.
(6) Choose Set | Grid Lines. The Grid Lines dialog box will be displayed.
(7) Select the At Major Ticks box.
(8) Click on OK.

Adding a Graph Title

(1) Choose Draw | Text. The normal cursor arrow will change to an arrow with a T.
(2) Position the cursor somewhere near the page coordinates X = 2, Y = 9. (The exact location is not necessary because the graph title may be repositioned at a later time.)
(3) Click on the chosen location. The Text dialog box will be displayed.
(4) Click on the arrow next to the Points edit line until the default number (e.g., 12) is replaced by a new font size (e.g., 24).

(5) In the text box at the bottom of the dialog box, type in the graph title (e.g., COOPER-JACOB GRAPH - Well No. 2) in the space following the blinking cursor bar.

(6) Click on OK.

(7) On the graph, select (i.e., click on) the graph title.

(8) Click and drag the title to a suitable location on the page (e.g., centered approximately 1 inch above the top of the graph).

(9) Unselect the graph title by clicking once outside of the plot frame.

Constructing a Best-Fit Line

(1) Select (i.e., click on) the plot line.

(2) Choose Set | Fits.

(3) Click on the arrow next to the best-fit equation (e.g., Linear, $Y = B*X + A$).

(4) Choose Log, $T = B*\log(X) + A$.

(5) Choose the Add button.

(6) Double click on the word Invisible on the Fit 1 text line.

(7) In the Style edit box, select the Single Solid Line button and click on OK.

(8) In the Plot Interval edit box, click on the Axis Limits button.

(9) In the Y-Axis Clipping check box, unselect the Y min.

(10) Click on OK. A Cooper-Jacob graph with a best-fit line will be displayed on the monitor screen.

Saving the Cooper-Jacob Graph

Save the graph/chart and data.

(1) Choose File | Save As. The Save As dialog box will be displayed.

(2) With the default file name (e.g., plotl.grf) selected, type in the new name for the hydrograph (e.g., cooper.grf).

(3) Click on OK. The data and the graph will be saved under the original file name.

Printing the Cooper-Jacob Graph

Before you print/plot, make sure that your application is properly configured for your particular printer. The instructions below provide only basic information about printing/plotting the graph. For more detailed instructions, see the application reference manual or click on the Help icon.

Check the printer/plotter configuration.

(1) Choose File | Change Printer. The Change Printer dialog box will appear on the screen.

(2) Select your printer/plotter from the Printer list box. (Use Setup to specify print settings.)

(3) Click on OK.

Print/plot the graph.

(1) Make sure that the graph (not the worksheet) is displayed and active.

(2) Choose File | Print.

(3) Click on OK.

Closing the Application and Quitting Windows

Before closing the application, be sure to save all active files.

(1) Choose File | Exit. If you have saved all active files, the application window closes. If you changed an active file but did not save it, then the Exit dialog box will appear.

(2) If the Exit dialog box appears, choose Yes to save the file. The Windows desktop will appear on the screen.

(3) Quit Windows and turn off the computer. Be sure to close all open applications before quitting Windows.

Calculating the Transmissivity and Storativity

Follow the procedure described in Section 22.2 to calculate transmissivity and storativity and record the resultant values on the Cooper-Jacob plot worksheet on Figure 22.3.

22.4 MICROCOMPUTER AND TYPE CURVE MATCHING SOFTWARE METHOD

Purpose

To estimate the transmissivity and storativity of an ideal aquifer by the Cooper-Jacob straight-line method.

Reference

Duffield, G. M., and Rumbaugh, J. O., III, 1989. *AQTESOLV: Aquifer Test Solver, Version 1.00, Documentation* (Version 1.10 update). Geraghty & Miller, Inc.

Equipment and Materials

- aquifer test data (e.g., Table 21.1)
- microcomputer with an 80386 processor or higher, MS-DOS 3.3 or higher, Windows 3.1 or higher, hard drive, floppy disk drive, and HP laser printer or plotter
- AQTESOLV for DOS (Version 1.10) software
- Hydrodata Diskette

Procedure

This method employs AQTESOLV to estimate aquifer properties and to construct a Cooper-Jacob graph from pumping test data. The progran can be used to create and save a file of aquifer test data (see Table 21.1 for an example, or use your own project data). Alternatively, if you have an existing AQTESOLV file of ideal aquifer data, it may be imported and used to construct a Cooper-Jacob graph.

Starting Up

AQTESOLV is DOS-based and can be run either from the DOS prompt (C:\>) or from the Windows *Run* command. (The Path command in your AUTOEXEC.BAT file should include AQTESOLV.)

(1) Turn on the computer, monitor, and plotter/printer. Wait until the Windows desktop is displayed.
(2) Choose File|Run, or click on the Windows 95 Start button and choose *Run*. The Run dialog box will appear.
(3) In the edit box, type AQTESOLV.
(4) Press ENTER. The AQTESOLV Program Control menu will be displayed on the screen.
(5) Insert the Hydrodata Diskette into drive A (or B).

Specify disk drive and directory to be used to retrieve and save data files.

(1) Choose Misc. Utilities.
(2) Choose Change Current Working Directory. The Change Current Working Directory dialog box will be displayed.
(3) Type A:\ (or B:\).
(4) Press ENTER ESC.

Creating an Aquifer Test Data File

If you are going to use a previously created AQTESOLV file of aquifer test data, you may skip this section of instructions and go on to *Loading an Aquifer Test Data File.*

Initialize a new data set.

(1) Choose Data Set Manager.

(2) Choose Create New Data Set. The Input Options menu will be displayed on the screen *Message:* Are you sure you want to initialize the data set? (Y/N): No.

(3) Type Y ENTER.

(4) Choose Title for Data Set. *Message:* Enter title:

(5) Type the data set title (e.g., IDEAL AQUIFER ANALYSIS).

(6) Press ENTER.

Specify the aquifer test data.

(1) Choose Confined/Leaky Aquifer Test Data. The Confined/Leaky Aquifer Test Data menu will be displayed on the screen.

(2) Type the pumping rate (e.g., 26.2).

(3) Press ENTER.

(4) Type the distance to observation well (e.g., 400).

(5) Press ENTER ESC. The Confined/Leaky Aquifer Test Data menu will disappear from the screen.

(6) Choose Observation Well Measurements. The Time–Drawdown Entry menu will be displayed on the screen.

(7) Choose Enter Data from Keyboard. The observation well data worksheet will be displayed on the screen.

(8) Type the first time measurement (e.g., 3).

(9) Press ENTER.

(10) Type the first drawdown measurement (e.g., 1.8).

(11) Press ENTER ENTER.

(12) Type the second time measurement (e.g., 4).

(13) Press ENTER.

(14) Type the second drawdown measurement (e.g., 2.3).

(15) Press ENTER ENTER. Repeat this procedure until all time–drawdown pairs have been entered. Check the entry data against the original measurements and correct any errors.

(16) Type ESC ESC ESC. *Message:* Save AQTESOLV data set on disk? (Y/N): Yes.

Saving the Aquifer Test Data File

(1) Press ENTER. *Message:* Enter file name for data set:

(2) Type the new data set filename (e.g., ideal.dat).

(3) Press ENTER. *Message:* Open new output file on disk? (Y/N): Yes.

(4) Press ENTER. *Message:* Enter output filename:

(5) Type the new output filename (e.g., cooper.out).

(6) Press ENTER ESC.

Skip the next section and go to *Analyzing the Data Set.*

Loading an Aquifer Test Data File

(1) Choose Data Set Manager.

(2) Choose Read Data Set from Disk. *Message:* Enter data set directory:

(3) Type the disk drive\directory [e.g., A:\ (or B:\)].

(4) Press ENTER. *Message:* Enter data set extension: *.dat.

(5) Press ENTER. *Message:* Enter data set filename:

(6) Type the data set filename (e.g., ideal.dat).

(7) Press ENTER. The time–drawdown data set will be displayed on the screen.

(8) Press ENTER ENTER. *Message:* Open new output file on disk? (Y/N): Yes.

(9) Press ENTER.

(10) Type the output filename (e.g., cooper.out).

(11) Press ENTER ESC. The Program Control menu will be displayed on the monitor screen.

Analyzing the Data Set

(1) Choose Confined Solutions. The Confined Solutions menu will be displayed on the screen.

(2) Choose 2 Cooper-Jacob Method. The Parameter Estimation Control menu will be displayed on the screen.

(3) Choose Estimate Aquifer Parameters. The program will automatically adjust aquifer coefficients to match the Cooper-Jacob plot to your time–drawdown data and display the values. *Message:* Perform additional nonlinear iterations? (Y/N): Yes.

(4) Type N ENTER.

(5) Choose Graph Results│Plot Graph. A Cooper-Jacob plot will be displayed on the screen. The estimates of transmissivity, T, and storativity, S, are printed in the upper-left corner of the graph. If time values have been entered in minutes, then transmissivity will be calculated in ft.^{2}/min.

Adjusting the Type Curve (optional)

Unless the Cooper-Jacob plot line on the screen makes an unsatisfactory match with the data points, do not modify the position of the plot line. If the plot line does not match the data points well, then refer to the AQTE-SOLV reference manual for assistance in moving the line to a more satisfactory position.

Printing/Plotting the Cooper-Jacob Graph

The instructions below assume that AQTESOLV is configured correctly for your particular printer or plotter. (If you need assistance, see the AQTESOLV documentation manual for configuring your hardware correctly.) You can output the Cooper-Jacob graph to either an HP plotter or a printer. Choose the appropriate option below.

Options (Select one)

Plot graph on an HP plotter.

(1) Place a black pen in pen position #1 and insert a sheet of blank paper.
(2) Press ALT-P. *Message:* Is your plotter connected? (Y/N): Yes.
(3) Press ENTER.

Print graph on a printer.

(1) Press ALT-L. *Message:* Sending plot to a printer. Strike a key (ESC = abort).
(2) Press ENTER. After a short wait, a Cooper-Jacob graph curve will be plotted/printed.

Saving the Cooper-Jacob Graph on Disk

(1) Press ESC ESC.
(2) Choose Print Results to Disk. The computer will beep twice when the graph has been saved on the disk.

Quitting the AQTESOLV Program

(1) Press F2 or choose Quit AQTESOLV from the Program Control menu. *Message:* Quit AQTESOLV? (Y/N): No.
(2) Type Y ENTER. The AQTESOLV program will disappear from the screen. If you ran the program from Windows, the Windows desktop will appear on the screen.

(3) Open another application or quit Windows and turn off the computer. Be sure to close all open applications before quitting Windows.

22.5 REFERENCES

Driscoll, F. G., 1986. *Groundwater and Wells* (2nd ed.). St. Paul, MN: Johnson Division.

Duffield, G. M., and Rumbaugh, J. O., III, 1989. *AQTESOLV : Aquifer Test Solver, Version 1.00, Documentation* (Version 1.10 update). Geraghty & Miller, Inc.

Heath, R. C., 1987. *Basic Ground-Water Hydrology.* U.S. Geological Survey Water-Supply Paper 2220.

Theis, C. V., 1935. The relation between the lowering of the piezometric surface and the rate and duration of discharge of a well using ground water storage. *Am. Geophysical Union Trans.,* pp. 518–525.

Analytical Procedure for Estimating Hydraulic Properties of Leaky Aquifers: Hantush-Jacob Method

23.1 INTRODUCTION TO METHODS

As discussed in Chapter 16, an aquifer bounded by confining beds that leak water into the aquifer in response to withdrawals is called a leaky aquifer, even though it is the confining bed that is leaky (see Figure 16.14). Under leaky conditions drawdowns differ from those that would be predicted by the Theis equation (see Figure 23.1).

Methods for analyzing so-called leaky aquifers are based chiefly on the works of Hantush and Jacob (1955) and Hantush (1960). The analytical solution of Hantush and Jacob is commonly stated in the same form as the Theis solution but with a more complicated well function. The particular equation employed depends on whether or not there is storage in the confining beds.

For leaky aquifers with no storage in confining beds, the type curves are constructed by plotting values of $W(u,r/B)$ against values of $1/u$ on logarithmic graph paper (see Figure 23.2). As in Theis type curve matching, time–drawdown data are plotted on logarithmic graph paper of the same scale as the type curves. The time–drawdown plot is then superimposed on the type curve graph and, keeping the axes parallel, adjusted until the data plot matches one of the type curves. A match point of the coordinates $W(u,r/B)$, $1/u$, s, and t is selected and marked on the data plot. These values are then substituted into the Hantush-Jacob equations to determine transmissivity and storativity.

Transmissivity is calculated from

$$T = \frac{Q}{4\pi s} W(u, r/R) \qquad (23.1)$$

273

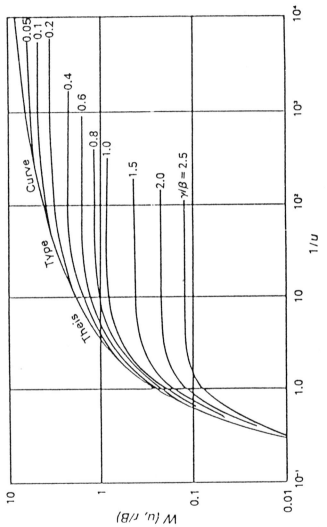

Figure 23.1 Hantush-Jacob curves of leaky aquifer with no storage in confining layer (from Heath, 1987, p. 51).

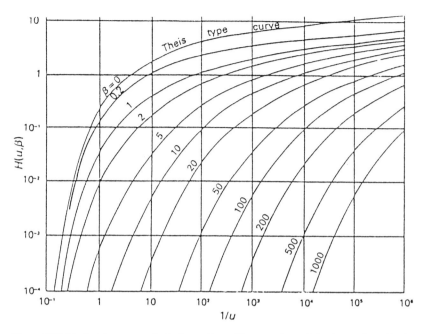

Figure 23.2 Hantush type curves for leaky aquifers with storage in confining layers (from Heath, 1987, p. 51).

where T is the transmissivity $[L^2/T]$, Q is the pumping rate $[L^3/T]$, s is the drawdown $[L]$, $W(u,r/B)$ is the leaky well function, $u = r^2S/4Tt$, $B = (Tb'/K')^{1/2}$, b' is the thickness of the leaky (confining) layer $[L]$, K' is the vertical hydraulic conductivity of the leaky layer $[L/T]$, r is the distance $[L]$ from pumped well to observation well, S is the storativity [dimensionless], defined below, and t is the time since pumping started $[T]$, storativity is calculated from the equation

$$S = \frac{4Ttu}{r^2} \qquad (23.2)$$

where the variables were previously defined.

For leaky aquifers with storage in confining beds, a different set of type curves is used (see Figure 23.3). Transmissivity is calculated from

$$T = \frac{Q}{4\pi s}H(u,r/B) \qquad (23.3)$$

where H is the well function for leaky aquifers with storage. [For additional information, see Walton (1970).]

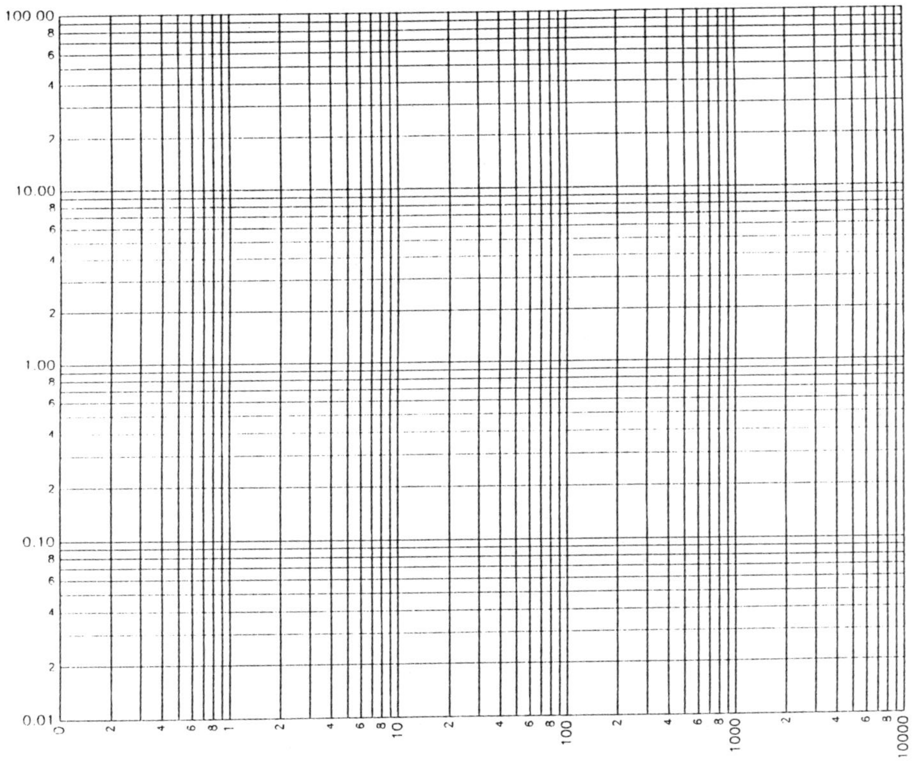

Figure 23.3 Logarithmic graph form for Hantush-Jacob plot.

276

TABLE 23.1. Aquifer Test Data for a Leaky Aquifer
(Q = 26 gpm,[1] r = 96 ft.)

	Drawdown	
Time (min.)		Drawdown (ft.)
5		0.76
28		3.30
41		3.59
60		4.08
75		4.39
244		5.47
493		5.96
669		6.11
959		6.27
1129		6.40
1885		6.42

[1]Before beginning calculations and plotting, convert Q from gpm to ft.^{3}/min.

This chapter describes two methods by which Hantush-Jacob type curves can be used to estimate the hydraulic properties of a leaky, confined aquifer with no storage in confining beds. They are the *Hand Plotting and Curve Matching Method* (Section 23.2) and the *Microcomputer and Type Curve Matching Software Method* (Section 23.3). The second method (Section 23.3) also permits the analysis of a leaky aquifer with storage in confining beds.

Table 23.1 displays time–drawdown data for a leaky aquifer. Use these data (or your own project data) for application of the methods described in this section. Before commencing calculation and plotting, convert Q from gpm to ft.^{3}/min.

23.2 HAND PLOTTING AND CURVE-MATCHING METHOD

Purpose

To construct a logarithmic graph of time–drawdown data, match the plotted data with Hantush-Jacob type curves, and estimate the transmissivity and storativity of a leaky aquifer with no storage in confining beds.

References

Heath, R. C., 1987. *Basic Ground-Water Hydrology.* U.S. Geological Survey Water-Supply Paper 2220, pp. 50–51.

Walton, W. C., 1970. *Groundwater Resource Evaluation,* New York, NY: McGraw-Hill Book Co., pp. 217–219.

Equipment and Materials

- aquifer test data (e.g., Table 23.1)
- logarithmic graph paper (e.g., Figure 23.3)
- Hantush-Jacob type curves (e.g., Figure 23.4)
- Hantush-Jacob type curve match worksheet (Figure 23.5)
- pencil (3H)

Procedure

Label field plot graph axes. On a sheet of logarithmic graph paper (e.g., Figure 23.3), label the bottom (horizontal) axis "TIME SINCE PUMPING STARTED (min)" and the left (vertical) axis "DRAWDOWN (ft)." If necessary, renumber the graph axes to accommodate the range of time and drawdown data.

Plot time–drawdown data. Using aquifer test data (e.g., Table 23.1, or your own project data, plot time–drawdown data pairs on logarithmic graph paper labeled for a field plot (e.g., Figure 23.3).

Superimpose a Hantush-Jacob graph on plot of field data. On a light table (or in lieu of this, a windowpane) superimpose the time–drawdown plot on a Hantush-Jacob type curve (e.g., Figure 23.4).

Select a match point.

(1) Determine a match point for the data plot and the type curve.
(2) Record the values of the match point on the Hantush-Jacob type curve match worksheet (Figure 23.5).

Calculate transmissivity.

(1) Enter the values for Q, t, $W(u,r/B)$, and s on the appropriate lines on the worksheet.
(2) Use the equation to calculate the transmissivity of the aquifer and record the values on the worksheet.

Calculate storativity.

(1) Enter the values of u and r on the appropriate lines of the worksheet and calculate r^2.

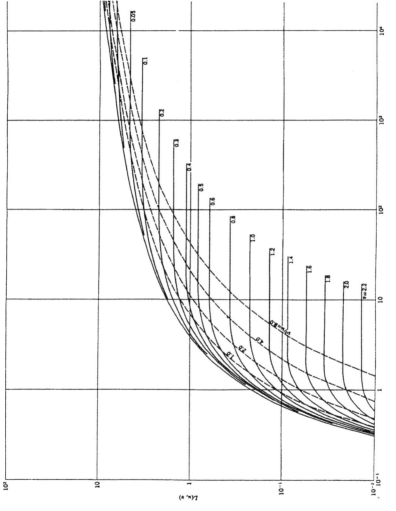

Figure 23.4 Hantush-Jacob type curves.

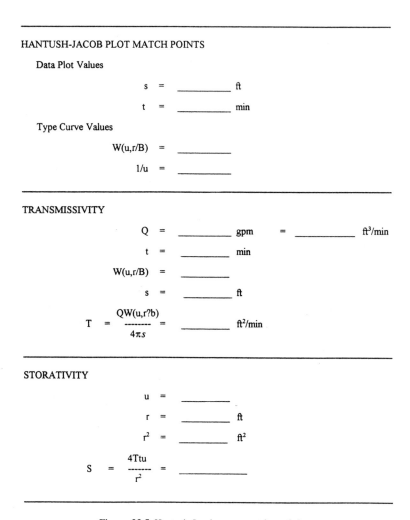

HANTUSH-JACOB PLOT MATCH POINTS

Data Plot Values

$$s = \text{_____} \text{ft}$$

$$t = \text{_____} \text{min}$$

Type Curve Values

$$W(u,r/B) = \text{_____}$$

$$1/u = \text{_____}$$

TRANSMISSIVITY

$$Q = \text{_____} \text{gpm} = \text{_____} \text{ft}^3/\text{min}$$

$$t = \text{_____} \text{min}$$

$$W(u,r/B) = \text{_____}$$

$$s = \text{_____} \text{ft}$$

$$T = \frac{QW(u,r?b)}{4\pi s} = \text{_____} \text{ft}^2/\text{min}$$

STORATIVITY

$$u = \text{_____}$$

$$r = \text{_____} \text{ft}$$

$$r^2 = \text{_____} \text{ft}^2$$

$$S = \frac{4Ttu}{r^2} = \text{_____}$$

Figure 23.5 Hantush-Jacob curve match worksheet.

(2) Use the equation to calculate the storativity of the aquifer and record on the worksheet.

23.3 MICROCOMPUTER AND TYPE CURVE MATCHING SOFTWARE METHOD

Purpose

To estimate the transmissivity and storativity of a leaky aquifer.

Reference

Duffield, G. M., and Rumbaugh, J. O., III, 1989. *AQTESOLV: Aquifer Test Solver, Version 1.00, Documentation.* (Version 1.10 update). Geraghty & Miller, Inc.

Equipment and Materials

- aquifer test data (e.g., Table 23.1)
- microcomputer with an 80386 processor or higher, MS-DOS 3.3 or higher, Windows 3.1 or higher, hard drive, floppy disk drive, and HP laser printer or plotter
- AQTESOLV for DOS (Version 1.10) software
- Hydrodata Diskette

Procedure

This method employs AQTESOLV to estimate aquifer properties and to construct a Hantush-Jacob graph from aquifer test data. The program can be used to create and save a file of aquifer test data (see Table 23.1 for an example, or use your own project data). Alternatively, if you have an existing AQTESOLV file of aquifer test data (e.g., leaky.dat), it may be imported and used to construct a Hantush-Jacob graph.

Starting Up

AQTESOLV is DOS-based and can be run either from the DOS prompt (C:\>) or from the Windows *Run* command. (The Path command in your AUTOEXEC.BAT file should include AQTESOLV.)

(1) Turn on the computer, monitor, and plotter/printer. Wait until the Windows desktop is displayed.

(2) Choose File | Run, or click on the Windows 95 Start button and choose *Run*. The Run dialog box will appear.

(3) In the edit box, type AQTESOLV.

(4) Press ENTER. The AQTESOLV Program Control menu will be displayed on the screen.

(5) Insert the Hydrodata Diskette into drive A (or B).

Specify disk drive and directory to be used to retrieve and save data files.

(1) Choose Misc. Utilities.

(2) Choose Change Current Working Directory. The Change Current Working Directory dialog box will be displayed.

(3) Type A:\ (or B:\).

(4) Press ENTER ESC.

Creating an Aquifer Test Data File

If you are going to use a previously created AQTESOLV file of aquifer test data, you may skip this section of instructions and go on to *Loading an Aquifer Test Data File.*

Initialize a new data set.

(1) Choose Data Set Manager.

(2) Choose Create New Data Set. The Input Options menu will be displayed on the screen. *Message:* Are you sure you want to initialize the data set? (Y/N): No.

(3) Type Y ENTER.

(4) Choose Title for Data Set. *Message:* Enter title:

(5) Type the data set title (e.g., LEAKY AQUIFER ANALYSIS).

(6) Press ENTER.

Specify the aquifer test data.

(1) Choose Confined/Leaky Aquifer Test Data. The Confined/Leaky Aquifer Test Data menu will be displayed on the screen.

(2) Type the pumping rate (e.g., 3.5).

(3) Press ENTER.

(4) Type the distance to the observation well (e.g., 96).

(5) Press ENTER ESC. The Confined/Leaky Aquifer Test Data menu will disappear from the screen.

(6) Choose Observation Well Measurements. The Time–Drawdown Entry menu will be displayed on the screen.

(7) Choose Enter Data from Keyboard. The observation well data worksheet will be displayed on the screen.

(8) Type the first time measurement (e.g., 5).

(9) Press ENTER.

(10) Type the first drawdown measurement (e.g., 0.76).

(11) Press ENTER ENTER.

(12) Type the second time measurement (e.g., 28).

(13) Press ENTER.

(14) Type the second drawdown measurement (e.g., 3.3).

(15) Press ENTER ENTER. Repeat this procedure until all time–drawdown pairs have been entered. Check the entry data against the original measurements and correct any errors.

(16) Type ESC ESC ESC. *Message:* Save AQTESOLV data set on disk? (Y/N): Yes.

Saving the Aquifer Test Data File

(1) Press ENTER. *Message:* Enter file name for data set:

(2) Type the new data set filename (e.g., leaky.dat).

(3) Press ENTER. *Message:* Open new output file on disk? (Y/N): Yes.

(4) Press ENTER. *Message:* Enter output filename:

(5) Type the new output filename (e.g., hantush.out).

(6) Press ENTER ESC.

Skip the next section and go on to *Analyzing the Data Set.*

Loading an Aquifer Test Data File

(1) Choose Data Set Manager.

(2) Choose Read Data Set from Disk. *Message:* Enter data set directory:

(3) Type the disk drive\directory [e.g., A:\ (or B:\)].

(4) Press ENTER. *Message:* Enter data set extension: *.dat.

(5) Press ENTER. *Message:* Enter data set filename:

(6) Type the data set filename (e.g., leaky.dat).

(7) Press ENTER. The time–drawdown data set will be displayed on the screen.

(8) Press ENTER ENTER. *Message:* Open new output file on disk? (Y/N): Yes.

(9) Press ENTER.

(10) Type the output filename (e.g., hantush.out).
(11) Press ENTER ESC. The Program Control menu will be displayed on the monitor screen.

Analyzing the Data Set

(1) Choose Leaky solutions. The Leaky Solutions menu will be displayed on the screen.
(2) Choose Hantush method, no storage in aquitards. The Parameter Estimation Control menu will be displayed on the screen.
(3) Choose Estimate Aquifer Parameters. The program will automatically adjust aquifer coefficients to match the Hantush-Jacob type curve to your time–drawdown data and display the values. *Message:* Perform additional nonlinear iterations? (Y/N): Yes.
(4) Type N ENTER.
(5) Choose Graph results | Plot Graph. A Hantush-Jacob type curve and field plot data will be displayed on the screen. The estimates of transmissivity, T, and storativity, S, are printed in the upper-left corner of the graph. If time values have been entered in minutes, then transmissivity will be calculated in ft.^{2}/min.

Adjusting the Type Curve (optional)

If necessary, you may change the position of the Hantush curve on the screen in order to obtain the "best" match with the data points by moving the mouse. The values of T and S will change automatically to reflect the new curve match. After a match has been achieved, refresh the screen and replot the graph with the new curve match by pressing F4.

Printing/Plotting the Hantush-Jacob Graph

The instructions below assume that AQTESOLV is configured correctly for your particular printer or plotter. (If you need assistance, see the AQTESOLV documentation manual for configuring your hardware correctly.) You can output the Hantush-Jacob graph to either an HP plotter or a printer. Choose the appropriate option below.

Options (select one)

Plot graph on an HP plotter.
(1) Place a black pen in pen position #1 and insert a sheet of blank paper.
(2) Press ALT-P. *Message:* Is your plotter connected? (Y/N): Yes.
(3) Press ENTER.

Print graph on a printer.

(1) Press ALT-L. *Message:* Sending plot to a printer. Strike a key (ESC = abort).

(2) Press ENTER. After a short wait, a Hantush-Jacob curve will be plotted/printed.

Saving the Hantush-Jacob Graph on Disk

(1) Press ESC ESC.

(2) Choose Print Results to Disk. The computer will beep twice when the graph has been saved on the disk.

Quitting the AQTESOLV Program

(1) Press F2 or choose Quit AQTESOLV from the Program Control menu. *Message:* Quit AQTESOLV? (Y/N): No.

(2) Type Y ENTER. The AQTESOLV program will disappear from the screen. If you ran the program from Windows, the Windows desktop will appear on the screen.

(3) Open another application or quit Windows and turn off the computer. Be sure to close all open applications before quitting Windows.

23.4 REFERENCES

Duffield, G. M., and Rumbaugh, J. O., III, 1989. *AQTESOLV: Aquifer Test Solver, Version 1.00, Documentation* (Version 1.10 update). Geraghty & Miller, Inc.

Freeze, R. A., and Cherry, J. A., 1979. *Groundwater.* Englewood Cliffs, NJ: Prentice-Hall, Inc.

Hantush, M. S., 1960. Modification of the theory of leaky aquifers. *Jour. of Geophys. Res.,* 65(11):3713–3725.

Hantush, M. S., and Jacob, C. E., 1955. Non-steady radial flow in an infinite aquifer. *Am. Geophys. Union Trans.,* 36:95–100.

Heath, R. C., 1987. *Basic Ground-Water Hydrology.* U.S. Geological Survey Water-Supply Paper 2220.

Walton, W. C., 1970. *Groundwater Resource Evaluation.* New York, NY: McGraw-Hill Book Co.

CHAPTER 24

Analytical Procedure for Estimating Hydraulic Properties of Unconfined Aquifers: Neuman Method

24.1 INTRODUCTION TO METHODS

W HEN water is pumped from an unconfined (water table) aquifer, the creation of a drawdown cone involves vertical (as well as horizontal) components of flow. As the water table begins to decline, water is derived from gravity drainage of the aquifer (see Figure 24.1). A result is that time–drawdown plots of unconfined aquifers deviate from the ideal (see Figure 24.2).

The simplest approach for analyzing the properties of unconfined aquifers is to use the Theis equation with the well function defined as specific yield, S_y, rather than storativity, S. Transmissivity, T, then is defined as $T = Kb$, where b is the saturated thickness of the aquifer before pumping began. As Walton (1970, p. 325) states, "Jacob (1950) has shown that this approach leads to predicted drawdowns that are very nearly correct as long as the drawdown is small in comparison with the saturated thickness."

The common, and more theoretically sound, method for analyzing unconfined aquifers is that developed by Neuman (1975). The analytical solution of Neuman is commonly stated in the same form as the Theis solution but with a more complicated well function. This method takes into account the existence of vertical flow and its evidential relationship to radius, r. The Neuman type curves are constructed by plotting values of $W(u_A, u_B, n)$ against values of $1/u_A$ and $1/u_B$ on logarithmic paper (see Figure 24.3). As in Theis type curve matching, time–drawdown data are plotted on logarithmic graph paper of the same scale as the type curves. The time–drawdown plot is then superimposed on the type curve graph and, keeping the axes parallel, adjusted until the plot of late data matches one

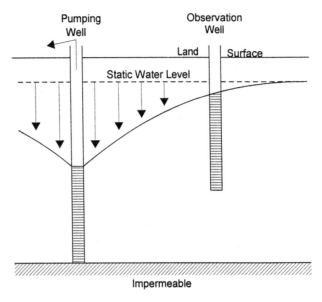

Figure 24.1 Illustration of a water table aquifer. Arrows denote gravity drainage of aquifer.

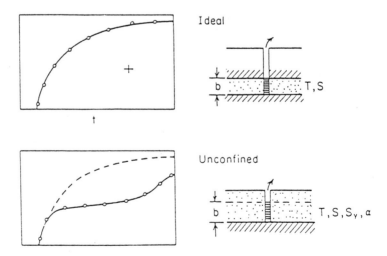

Figure 24.2 Difference between the type curves of an ideal aquifer and an unconfined aquifer (adapted from Freeze and Cherry, 1979, p. 346).

288

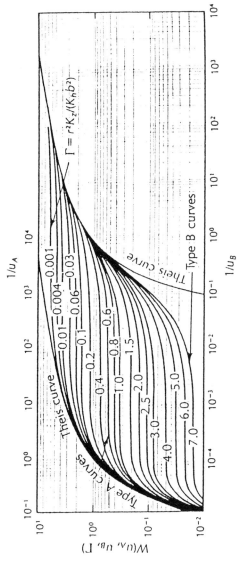

Figure 24.3 Illustration of Neuman type curves for unconfined aquifers (reprinted from Fetter, 1994, p. 238; adapted by permission of Prentice-Hall, Upper Saddle River, NJ).

of the Type-B curves. A match point of the coordinates $W(u_B)$, $1/u_B$, s, and t is selected and marked on the data plot. Then the early time–drawdown data are superimposed on a Type-A curve for the n-value of the previously matched Type-B curve and a second match point is determined for $W(u_A)$, $1/u_A$, s, and t. These values are then substituted into the Neuman equations to determine transmissivity, specific yield, and storativity. Additionally, the equations can be employed to determine both vertical hydraulic conductivity (K_v) and horizontal hydraulic conductivity (K_h).

Using the Neuman solution, transmissivity is calculated from

$$T = \frac{Q}{4\pi s} W(u_A, u_B, n) \qquad (24.1)$$

where $W(u_A, u_B, n)$ is the well function for the unconfined aquifer,

$$u_A = \frac{r^2 S}{4Tt} \qquad (24.2)$$

for early drawdown data

$$u_B = \frac{r^2 S_y}{4Tt} \qquad (24.3)$$

for later drawdown data

$$n = \frac{r^2 K_z}{b^2 K_b} \qquad (24.4)$$

T is the transmissivity [L^2/T], Q is the pumping rate [L^3/T], s is the drawdown [L], b is the initial saturated thickness [L], K_h is the horizontal hydraulic conductivity of the leaky layer [L/T], K_z is the vertical hydraulic conductivity of the leaky layer [L/T], r is the distance [L] from pumped well to observation well, S_y is specific yield [dimensionless], t is the time since pumping started [T], S is the storativity [dimensionless], defined as

$$S = \frac{4Ttu}{r^2} \qquad (24.5)$$

[For additional information, see Walton (1970).]

The methods described in this chapter are the *Hand Plotting and Curve Matching Method* (Section 24.2) and the *Microcomputer and Type Curve Matching Software Method* (Section 24.3).

TABLE 24.1. Aquifer Test Data for an Unconfined
Aquifer (Q = 144.4 ft.^{3}/min., r = 73 ft., b = 78 ft.).

Time (min.)	Drawdown (ft.)	Time (min.)	Drawdown (ft.)
1	0.64	100	1.31
2	0.86	120	1.36
3	0.94	150	1.45
4	0.97	200	1.52
5	0.98	250	1.59
6	0.99	300	1.65
8	1.01	350	1.70
10	1.02	400	1.75
15	1.03	500	1.85
20	1.06	600	1.95
25	1.08	700	2.01
30	1.13	800	2.09
35	1.15	900	2.15
40	1.17	1000	2.20
50	1.19	1200	2.27
60	1.22	1500	2.35
70	1.25	2000	2.49
80	1.28	2500	2.59
90	1.29	3000	2.66

Table 24.1 displays aquifer test data. Use these data (or your own project data) for application of the methods described in this chapter.

24.2 HAND PLOTTING AND CURVE MATCHING METHOD

Purpose

To construct a logarithmic graph of aquifer test data, match this fieldplot with a Neuman type curve, and calculate the transmissivity and storativity of an aquifer.

Reference

Walton, W. C., 1970. *Groundwater Resource Evaluation*. New York, NY: McGraw-Hill Book Co., pp. 222–224.

Equipment and Materials

- aquifer test data (e.g., Table 24.1)

- logarithmic graph paper (e.g., Figure 24.4)
- Neuman type curves (e.g., Figure 24.5)
- Neuman type curve match worksheet (Figure 24.6)
- pencil (3H)

Procedure

Label field plot of data. On a sheet of logarithmic graph paper (e.g., Figure 24.4), label the bottom (horizontal) axis "TIME SINCE PUMP-ING STARTED (min)" and the left (vertical) axis "DRAWDOWN (ft)." If necessary, renumber the graph axes to accommodate the range of time and drawdown data.

Plot time–drawdown data. Using time–drawdown data (e.g., Table 24.1 or your own project data), plot time–drawdown data pairs on logarithmic graph paper labeled for a fieldplot (e.g., Figure 24.4).

Superimpose the Neuman graph on the plot of field data.

(1) On a light table (or in lieu of this, a windowpane) superimpose the time–drawdown plot on the Neuman type curves (Figure 24.5).

(2) Keeping the graph axes parallel, adjust the Type-B curve until it matches the late time–drawdown data plot.

Select match point for Type-A curve.

(1) Determine a match point of the coordinates $W(u_B)$, $1/u_B$, s, and t.

(2) Record the values of the match point on Figure 24.6.

(3) Repeat the procedure matching the Type-A curve to the early time–drawdown data plot.

Select match point for Type-B curve.

(1) Determine a match point of the coordinates $W(u_A)$, $1/u_A$, s, and t.

(2) Record the values of the match point on the Neuman type curve match worksheet (Figure 24.6).

Calculate transmissivity. Calculate the transmissivity, T, of the aquifer and record the values on the Neuman type curve match worksheet.

Calculate storativity.

(1) Enter the values of u and r on the appropriate lines on the worksheet.

(2) Calculate r^2 and record on the worksheet.

(3) Calculate the storativity, S, of the aquifer and record on the work-sheet.

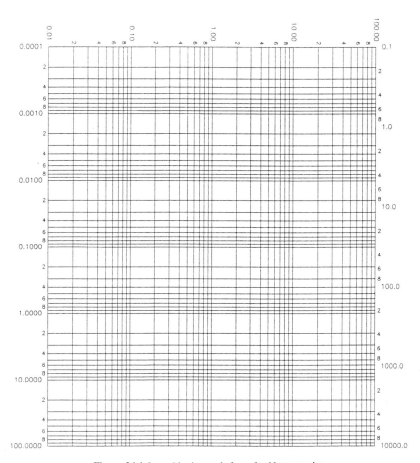

Figure 24.4 Logarithmic graph form for Neuman plot.

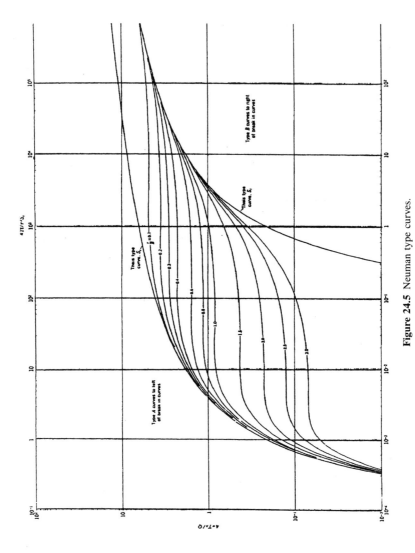

Figure 24.5 Neuman type curves.

294

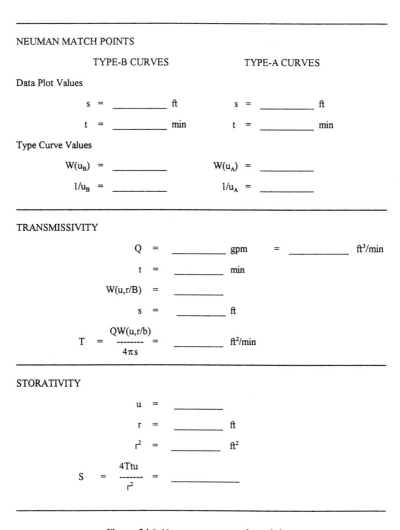

NEUMAN MATCH POINTS

TYPE-B CURVES   TYPE-A CURVES

Data Plot Values

s = _____ ft s = _____ ft

t = _____ min t = _____ min

Type Curve Values

$W(u_B)$ = _____ $W(u_A)$ = _____

$1/u_B$ = _____ $1/u_A$ = _____

TRANSMISSIVITY

Q = _____ gpm   = _____ ft³/min

t = _____ min

$W(u,r/B)$ = _____

s = _____ ft

$T = \dfrac{QW(u,r/b)}{4\pi s}$ = _____ ft²/min

STORATIVITY

u = _____

r = _____ ft

r^2 = _____ ft²

$S = \dfrac{4Ttu}{r^2}$ = _____

Figure 24.6 Neuman curve match worksheet.

24.3 MICROCOMPUTER AND TYPE CURVE MATCHING SOFTWARE METHOD

Purpose

To estimate the transmissivity and storativity of an unconfined (water table) aquifer.

Reference

Duffield, G. M., and Rumbaugh, J. O., III, 1989. *AQTESOLV: Aquifer Test Solver, Version 1.00, Documentation* (Version 1.10 update). Geraghty & Miller, Inc.

Equipment and Materials

- aquifer test data (e.g., Table 24.1)
- microcomputer with an 80386 processor or higher, MS-DOS 3.3 or higher, Windows 3.1 or higher, hard drive, floppy disk drive, and HP laser printer or plotter
- AQTESOLV for DOS (Version 1.10) software
- Hydrodata Diskette

Procedure

This method employs AQTESOLV to estimate aquifer properties and to construct a Neuman graph from aquifer test data. The program can be used to create and save a file of aquifer test data (see Table 21.1 for an example, or use your own project data). Alternatively, if you have an existing AQTESOLV file of aquifer test data (e.g., unconfin.dat), it may be imported and used to construct a Neuman graph.

Starting Up

AQTESOLV is DOS-based and can be run either from the DOS prompt (C:\>) or from the Windows *Run* command. (The Path command in your AUTOEXEC.BAT file should include AQTESOLV.)

(1) Turn on the computer, monitor, and plotter/printer. Wait until the Windows desktop is displayed.
(2) Choose File|Run, or click on the Windows 95 Start button and choose *Run*. The Run dialog box will appear.
(3) In the edit box, type AQTESOLV.

(4) Press ENTER. The AQTESOLV Program Control menu will be displayed on the screen.

(5) Insert the Hydrodata Diskette into drive A (or B).

Specify disk drive and directory to be used to retrieve and save data files.

(1) Choose Misc. Utilities.

(2) Choose Change Current Working Directory. The Change Current Working Directory dialog box will be displayed.

(3) Type A:\ (or B:\ ).

(4) Press ENTER ESC.

Creating an Aquifer Test Data File

If you are going to use a previously created AQTESOLV file of aquifer test data, you may skip this section of instructions and go on to *Loading an Aquifer Test Data File.*
Initialize a new data set.

(1) Choose Data Set Manager.

(2) Choose Create New Data Set. The Input Options menu will be displayed on the screen. *Message:* Are you sure you want to initialize the data set? (Y/N): No.

(3) Type Y ENTER.

(4) Choose Title for Data Set. *Message:* Enter title:

(5) Type the data set title (e.g., UNCONFINED AQUIFER ANALYSIS).

(6) Press ENTER.

Specify the aquifer test data.

(1) Choose Unconfined Aquifer Test Data. The Unconfined Aquifer Test Data menu will be displayed on the screen.

(2) Type the pumping rate (e.g., 134).

(3) Press ENTER.

(4) Type the distance to observation well (e.g., 200).

(5) Press ENTER.

(6) Type the saturated thickness (e.g., 100).

(7) Press ENTER. The Unconfined Aquifer Test Data menu will disappear from the screen.

(8) Choose Observation Well Measurements. The Time–Drawdown Entry menu will be displayed on the screen.

(9) Choose Enter Data from Keyboard. The observation well data worksheet will be displayed on the screen.

(10) Type the first time measurement (e.g., 1).

(11) Press ENTER.

(12) Type the first drawdown measurement (e.g., 0.09).

(13) Press ENTER ENTER.

(14) Type the second time measurement (e.g., 2).

(15) Press ENTER.

(16) Type the second drawdown measurement (e.g, 0.22).

(17) Press ENTER ENTER. Repeat this procedure until all time–drawdown pairs have been entered. Check the entry data against the original measurements and correct any errors.

(18) Press ESC ESC ESC. *Message:* Save AQTESOLV data set on disk? (Y/N): Yes.

Saving the Aquifer Test Data File

(1) Press ENTER. *Message:* Enter file name for data set:

(2) Type the new data set filename (e.g., unconfin.dat).

(3) Press ENTER. *Message:* Open new output file on disk? (Y/N): Yes.

(4) Press ENTER. *Message:* Enter output filename:

(5) Type the new output filename (e.g., neuman.out).

(6) Press ENTER ESC.

Skip the next section and go on to *Analyzing the Data Set.*

Loading an Aquifer Test Data File.

(1) Choose Data Set Manager.

(2) Choose Read Data Set from Disk. *Message:* Enter data set directory.

(3) Type the disk drive\directory [e.g., A:\ (or B:\)].

(4) Press ENTER. *Message:* Enter data set extension: *.dat.

(5) Press ENTER. *Message:* Enter data set filename:

(6) Type the data set filename (e.g., unconfin.dat).

(7) Press ENTER. The time–drawdown data set will be displayed on the screen.

(8) Press ENTER ENTER. *Message:* Open new output file on disk? (Y/N): Yes.

(9) Press ENTER.

(10) Type the output filename (e.g., neuman.out).

(11) Press ENTER ESC. The Program Control menu will be displayed on the monitor screen.

Analyzing the Data Set

(1) Choose Unconfined Solutions. The Unconfined Solutions menu will be displayed on the screen.

(2) Choose Neuman Method. The Parameter Estimation control menu will be displayed.

(3) Choose Graph Data/Match Curve.

(4) Choose Plot Graph. A preliminary data plot will appear on the screen.

Display and activate the "Type-A" curves.

(1) Press ALT-A. Three Type-A curves will be displayed on the monitor screen. The value of β displayed corresponds to the solid line. To increase or decrease the value of β, use the $+$ or $-$ keys. To move the solid Type-A curve, use the mouse.

(2) Match the Type-A curve to the early time–drawdown data by using the arrow keys.

Display and activate the "Type B" curves.

(1) Press ALT-B. Three Type-B curves will be displayed on the monitor screen. The value of β displayed corresponds to the solid line. To increase or decrease the value of β, use the $+$ or $-$ keys. To move the solid Type-B curve, use the mouse.

(2) Match the Type-B curve to the late time–drawdown data by using the mouse.

Display the complete Neuman curve for the plot.

(1) Press F4.

Return to Plot Options menu.

(1) Press ESC.

Create a graph of the results.

(1) Choose Graph Results | Plot Graph. A Neuman curve and field plot data will be displayed on the screen. The estimates of transmissivity, T, and storativity, S, are printed in the upper-left corner of the graph. To move the position of the type curve on the screen, use the arrow keys.

Printing/Plotting the Neuman Graph

The instructions below assume that AQTESOLV is configured correctly for your particular printer or plotter. (If you need assistance, see the AQTESOLV documentation manual for configuring your hardware correctly.) You can output the Neuman graph to either an HP plotter or a printer. Choose the appropriate option below.

Options (select one)

Plot graph on an HP plotter.

(1) Place a black pen in pen position #1 and insert a sheet of blank paper.

(2) Press ALT-P. *Message:* Is your plotter connected? (Y/N): Yes.

(3) Press ENTER.

Print graph on a printer.

(1) Press ALT-L. *Message:* Sending plot to a printer. Strike a key (ESC = abort).

(2) Press ENTER. After a short wait, a Neuman type curve will be plotted/printed.

Saving the Neuman Graph on Disk

(1) Press ESC ESC.

(2) Choose Print Results to Disk. The computer will beep twice when the graph has been saved on the disk.

Quitting the AQTESOLV Program

(1) Press F2 or choose Quit AQTESOLV from the Program Control menu. *Message:* Quit AQTESOLV? (Y/N): No.

(2) Type Y ENTER. The AQTESOLV program will disappear from the screen. If you ran the program from Windows, the Windows desktop will appear on the screen.

(3) Open another application or quit Windows and turn off the computer. Be sure to close all open applications before quitting Windows.

24.4 REFERENCES

Duffield, G. M., and Rumbaugh, J. O., III, 1989. *AQTESOLV: Aquifer Test Solver, Version 1.00, Documentation* (Version 1.10 update). Geraghty & Miller, Inc.

Fetter, C. W., Jr., 1994. *Applied Hydrogeology.* 3rd ed., New York, NY: Macmillan College Publ. Co.,

Freeze, R. A., and Cherry, J. A., 1979. *Groundwater.* Englewood Cliffs, NJ: Prentice-Hall, Inc.

Neuman, S. P., 1975. Analysis of pumping test data from anisotropic unconfined aquifers considering delayed yield. *Water Resources Research,* 11(2):329–342.

Walton, W. C., 1970. *Groundwater Resource Evaluation.* New York, NY: McGraw-Hill Book Co., Quoting Jacob, 1950. . .

Preparing a Hydrodata Diskette

IN order to carry out several of the microcomputer applications described in this manual, you must first prepare a floppy disk of special spreadsheet templates and graph data files, called the Hydrodata Diskette. The template files are electronic forms on which data and formulas are entered (see Sections 4.3, 8.3, 12.3, and 19.3). Each of these templates is illustrated in Figures A.1–A.4. The graph data files are employed to create the well yield frequency graphs that are described in Section 13.3.

CREATING SPREADSHEET TEMPLATES

(1) Format a high-density 3.5-inch floppy diskette and label it HYDRODATA.

(2) Leave the newly formatted diskette in the floppy disk drive and open a Windows-based spreadsheet application of your choice (Quattro Pro, Microsoft Excel, Lotus 1-2-3, etc.). A blank worksheet will be displayed in the application window.

(3) Reproducing exactly the format illustrated by one of the template examples (e.g., Figure A.1), type in all labels and numerical values. Be certain to enter each label and number in the precise cell shown in the example. Moreover, do not change the default width of the columns.

(4) When you have typed all labels and numbers into the worksheet, save the spreadsheet as a text (ASCII) file (e.g., levtemp.txt).

(5) Create other template files by repeating Steps 3 and 4 for each of the remaining forms illustrated by Figures A.2–A.4.

CREATING WELL YIELD FREQUENCY GRAPH FILES

(1) Open a Windows-based word processing application of your choice (e.g.,

A	A	B	C	D	E	F	G
1	levtemp.txt						
2							
3							
4				WATER LEVEL DATA			
5							
6	DATE:						
7							
8	MEASURED BY:						
9							
10	===						
11							
12	Enter depth data in						DEPTH T(
13	columns A & B				HEIGHT	ELEV OF	WATER L
14	DEPTH TO		WELL	TOP OF	OF WELL	WATER	BELOW L
15	WATER LEVEL		NUMBER	CASING	CASING	LEVEL	SURFACE
16	feet	inches		feet	feet	feet	feet
17							
18	===						
19							
20	1	1	P-1	522.06	2		
21	1	1	P-2	530.98	0.6		
22	1	1	P-3	523.99	2.2		
23	1	1	P-4	529.96	2.7		
24	1	1	P-5	530.25	2.4		
25	1	1	P-6	529.61	2.8		
26	1	1	P-7	523.88	4.4		
27	1	1	P-8	527.9	2.3		
28	1	1	P-9	523.22	2.6		
29	1	1	P-10	521.81	2.8		
30	1	1	P-11	535.1	2.1		
31	1	1	P-12	525.7	3.1		

Figure A.1. An illustration of the levtemp.txt template format.

A	A	B	C	D	E	F	G	H
1	RATE AND VELOCITY OF GROUNDWATER FLOW WORKSHEET							
2								
3	Parameter				Value		Units	
4	max. water level			h1	1		ft or m	
5	min. water level			h2	1		ft or m	
6	length of flow path			l	1		ft or m	
7	hydraulic conductivity			K	1		ft or m/day	
8	saturated thickness			b	1		ft or m	
9	width of aquifer section			w	1		ft or m	
10	porosity			n	1		dimensionless	
11								
12	HYDRAULIC GRADIENT			(h1-h2)/l	1		dimensionless	
13	DARCY VELOCITY			Vdf	1		ft or m/day	
14	RATE OF FLOW			Q	1		ft^3 or m^3/day	
15	FLOW VELOCITY			Va	1		ft or m/day	
16	TRAVEL TIME			t	1		days	

Figure A.2. An illustration of the flowtemp.txt template format.

A	A	B	C	D	E	F	G	H
1	yield.txt							
2					Distribution Calculations			
3	Enter well yield data							
4	in columns A & B						Cumu'tive	Percent
5	Well	Well	Class			Relative	Relative	Rank
6	Number	Yield	Interval		Freq'ncy	Freq'ncy	Freq'ncy	= or >
7			2		1	1	1	1
8			5		1	1	1	1
9			10		1	1	1	1
10			20		1	1	1	1
11			30		1	1	1	1
12			40		1	1	1	1
13			50		1	1	1	1
14			60		1	1	1	1
15			70		1	1	1	1
16			80		1	1	1	1
17			90		1	1	1	1
18			100		1	1	1	1
19					1	1	1	1
20			TOTAL		1	1		

Figure A.3. An illustration of the yield.txt template format.

	A	B	C	D	E	F	G	H	I	J
1	fplot.txt									
2						AQUIFER TEST DATA				
3										
4	Project:						Observation well no.:			
5										
6	Location:						Pumped well no.:			
7										
8										
9	AVER. Q		1 gpm			r =		1 ft		1 m
10										
11										
12	=		1 ft/min^3							
13						r^2 =		1 ft^2		1 m^2
14	=		1 ft/day^3							
15										
16	=		1 m/day^3							
17								Hour	Min	
18	STATIC WATER LEVEL =			1 ft		TIME PUMP ON =		1	1	
19										
20							DRAW-		DRAW-	
21				DEPTH		TIME	DOWN	TIME	DOWN	
22				TO WATER		t	s	t	s	
23	Date	Hour	Min	Feet		min	feet	days	meters	
24	MM-DD-YR									
25		1	1	1		1	1	1	1	
26		1	1	1		1	1	1	1	
27		1	1	1		1	1	1	1	
28		1	1	1		1	1	1	1	
29		1	1	1		1	1	1	1	
30		1	1	1		1	1	1	1	
31		1	1	1		1	1	1	1	
32		1	1	1		1	1	1	1	
33		1	1	1		1	1	1	1	
34		1	1	1		1	1	1	1	
35		1	1	1		1	1	1	1	
36		1	1	1		1	1	1	1	
37		1	1	1		1	1	1	1	
38		1	1	1		1	1	1	1	
39		1	1	1		1	1	1	1	
40		1	1	1		1	1	1	1	
41		1	1	1		1	1	1	1	
42		1	1	1		1	1	1	1	
43		1	1	1		1	1	1	1	
44		1	1	1		1	1	1	1	
45		1	1	1		1	1	1	1	
46		1	1	1		1	1	1	1	
47		1	1	1		1	1	1	1	
48		1	1	1		1	1	1	1	
49		1	1	1		1	1	1	1	
50		1	1	1		1	1	1	1	
51		1	1	1		1	1	1	1	
52		1	1	1		1	1	1	1	

Figure A.4. An illustration of the fplot.txt template format.

TABLE A.1. Well Yield Frequency Graph Data.

DC	DMH	DOO	DTR	SBM	SC	SKT	SWC
90, 5	100, 5	87, 10	60, 5	30, 5	75, 5	99, 12	87, 5
60, 15	35, 15	55, 15	29, 15	20, 15	35, 15	20, 50	48, 15
17, 50	15, 50	21, 50	11, 50	12, 50	14, 50	7, 85	19, 50
7, 85	5.5, 85	6, 85	4, 85	7, 85	6, 85	3, 95	9, 85
5, 95	3, 95	3.5, 95	2, 95	4, 95	3, 95		4, 95

Microsoft Word, Wordperfect, etc.). For each of the aquifers illustrated in Table A.1 (e.g., DC, DMH, DOO, etc.), you must create a separate file.

(2) Click on FILE | NEW. A blank page will be displayed in the application window.

(3) Select one of the aquifers on Table A.1 (e.g., DC) and, beginning at the upper left corner of the new page, type the top data pair of the column, so that the numbers are separated by a comma (e.g., 90, 5).

(4) Press <Enter> to move the cursor to the next line and type in the second data pair.

(5) Proceed down the column until all data pairs have been typed.

(6) Using the code for the aquifer (e.g., DC), save the data as a text (ASCII) file (e.g., dc.txt).

(7) Close the active file and open a new document window.

(8) Repeat the procedure for each of the aquifers listed in Table A.1.

(9) Keep the Hydrodata Diskette handy for use with the microcomputer applications of this manual.